SpringerBriefs in Quantitative Finance

Series Editors

Pauline Barrieu, London School of Economics, London, United Kingdom
Lorenzo Bergomi, Société Générale, Paris, France
Jakša Cvitanić, EDHEC Business School, Nice Cedex 3, France
Matheus Grasselli, The Fields Institute for Research in the Math Sciences, ON, Canada
Nizar Touzi, École Polytechnique, Palaiseau Cedex, France
Vladimir Piterbarg, Barclays Capital, London, United Kingdom

For further volumes:
http://www.springer.com/series/8784

Pablo Azcue • Nora Muler

Stochastic Optimization in Insurance

A Dynamic Programming Approach

Springer

Pablo Azcue
Department of Mathematics & Statistics
Universidad Torcuato Di Tella
Buenos Aires, Argentina

Nora Muler
Department of Mathematics & Statistics
Universidad Torcuato Di Tella
Buenos Aires, Argentina

ISSN 2192-7006 ISSN 2192-7014 (electronic)
ISBN 978-1-4939-0994-0 ISBN 978-1-4939-0995-7 (eBook)
DOI 10.1007/978-1-4939-0995-7
Springer New York Heidelberg Dordrecht London

Library of Congress Control Number: 2014940733

Mathematics Subject Classification (2010): 91B30, 97M30, 49L25, 93E20

JEL Classifications: G220, C610, D810

Printed on acid-free paper

Springer is part of Springer Science+Business Media (www.springer.com)

To María and Juan
To Dan, Abel and my parents

Preface

The classical collective risk model was introduced by Lundberg [46] in 1903 and developed by Crámer [20] in 1930 to describe the free surplus process of an insurance company. In this model, called Cramér–Lundberg, the premiums are assumed to be collected continuously over time with constant intensity and the total claim amount at a given time is given by a compound Poisson process. Initially, the main problem of classical risk theory was to calculate the probability of ruin, but De Finetti [24] proposed in 1957 a more realistic and economically motivated stability criterion: the management of the company should look for maximizing the expectation of the present value of all dividends paid to the shareholders up to ruin time.

Nowadays, the main problems of stochastic control in insurance are to minimize the ruin probability and to maximize cumulative expected discounted dividend payouts, where the insurer can control the risk in several ways. One possibility is to invest dynamically part of the surplus on financial assets. Another possibility is to pass part of the premium to the reinsurer, which in return covers certain fraction of the claims.

The usual approach to deal with these kinds of problems is the method of dynamic programming. This approach was introduced by Bellman [14] in 1954 for optimal deterministic control problems. The basic idea is to relate the optimal problem with a certain differential equation called the Hamilton–Jacobi–Bellman equation (HJB). If the solution of the HJB equation exists, then this solution would be the optimal value function of the original control problem. However, there could be a trouble in the classical dynamic programming approach: there has to be a classical solution to the HJB equation, i.e., the solution has to be smooth to the order of derivatives involved in the equation.

In the problems of maximizing the survival probability, the corresponding HJB equations have classical solutions when the claim-size distribution has bounded density, but this is not generally the case in the problems of maximizing dividends. However, this holds under certain conditions on the claim-size distributions; see the comments and remarks of Chap. 5. In the other cases the optimization problem cannot be solved in this framework. One of the reasons that makes these problems harder for a general claim-size distribution is that the associated HJB equations involve integrodifferential operators due to the jumps in the free surplus process. In order to overcome this issue, some authors studied the diffusion approximation to the Cramér–Lundberg model; this approximation simplifies the HJB equations: they are ordinary differential equations with classical solutions.

In order to solve these problems in the general setting, it is natural to consider a weaker definition of solutions of the HJB equation; the notion of viscosity solutions introduced by Crandall and Lions [21] in 1983 is especially well suited for this task. This tool enables to find solutions of first-order integrodifferential equations or degenerate second-order integrodifferential equations. This approach has been widely used in finance theory for a long time; however it was not used in insurance control until quite recently.

The aim of this brief is to address the problem of maximization of survival probability as well as the maximization of dividends in the classical collective risk model using the viscosity approach. First we will show that the optimal value function can be characterized as either the unique or the smallest viscosity solution of the associated HJB equation, and then we found a strategy (the optimal) whose value function coincides with the optimal value function. In the problem of maximizing the survival probability, we will show that both the optimal reinsurance and the optimal investment controls depend only on the current surplus. The same holds for the problem of optimal dividend payments and besides this, the optimal dividend strategy has a band structure; roughly speaking, this means that the payment of dividends depends only on the current surplus and it is characterized by three sets $\mathcal{A}$, $\mathcal{B}$, and $\mathcal{C}$ which partitioned the state space of the surplus process. Each of these sets is associated with a certain dividend payments action. The concept of band strategy was introduced by Gerber [29].

This work is organized as follows: In Chap. 1, we present the classical collective risk model for an insurance company and introduce the notion of the survival probability and the optimal expectation of the discounted dividend payments as functions of the initial surplus. We also study the basic properties of these value functions and derive the associated HJB equations. In Chap. 2, we introduce two ways to control the risk: reinsurance and investment. We study the basic properties of the survival probability functions as well as the optimal dividend payments with reinsurance and investment. We also derive the associated HJB equations in all these cases. In Chap. 3, we introduce the notion of viscosity solutions and show that the value functions are indeed viscosity solutions of the corresponding HJB equations. In Chap. 4, we characterize the optimal value functions among the viscosity solutions of the corresponding HJB equations. In Chap. 5, we show the existence of optimal stationary strategies and describe their structure. In Chap. 6, we present a method to construct systematically the optimal value functions and the optimal strategies in a quite general setting and show some numerical examples.

Buenos Aires, Argentina

Pablo Azcue
Nora Muler

Contents

Chapter 1
Stability Criteria for Insurance Companies

In this chapter we present the classical collective risk model for an insurance company and introduce two ways of measuring the stability of the company: survival probability and the maximization of the expectation of the discounted dividend payments. We consider these stability measures as functions of the initial surplus; they are called the *value functions* of the corresponding problems. We present here the bare case; in later chapters we will also allow the company to control the risk by means of reinsurance and investment.

We show first some basic properties of the value functions and then we derive heuristically the differential equations associated to them; both are first-order integrodifferential equations. The problem of optimal dividend payments is a control one and the associated equation is called the *Hamilton–Jacobi–Bellman equation* (HJB, for short).

We also introduce the diffusion approximation for the classical collective risk model and obtain the equations associated to both the survival probability and the optimal dividend payments. In this setting, they turn to be second-order differential equations.

Finally, we discuss the difficulties involved in solving the problems in both settings and how to deal with them. In later chapters, after introducing the necessary tools, we actually fully characterized the value functions.

1.1 The Classical Collective Risk Model

The *collective risk models* are called collective because the risks in the portfolio of the insurance company are seen as a whole. Consider a constant portfolio of clients, the insurance company gets a constant stream of income from premiums with rate p and uses the surplus to pay the claims. The sizes of the claims and the time points at which the claims occur are random variables. We call (τ_i, U_i) the time and the size of the ith claim. We define N_t as the numbers of claims up to time t, that is

P. Azcue and N. Muler, *Stochastic Optimization in Insurance: A Dynamic Programming Approach*, SpringerBriefs in Quantitative Finance, DOI 10.1007/978-1-4939-0995-7_1,

$$N_t = \max\{i : \tau_i \le t\}.$$

Given an initial surplus x, the free surplus X_t of the insurance company at time t can be written as

$$X_t = x + pt - \sum_{i=1}^{N_t} U_i. \tag{1.1}$$

We define the *ruin time* of the company as

$$\tau = \min\{t : X_t < 0\}. \tag{1.2}$$

A typical realization of the claim and the corresponding free surplus process are depicted in Fig. 1.1a, b. In the time intervals between two consecutive claims, the surplus grows with constant slope p and drops by U_i at the ith claim-arrival time. At time $\tau = \tau_7$ the surplus of the company becomes negative for the first time.

We make the following assumptions on the distribution of sizes and occurrences of the claims:

(1) The first claim cannot occur at time zero, two claims cannot occur at the same time, and the number of claims in any time interval is finite. So $0 < \tau_1 < \tau_2 < \tau_3 < \cdots$, $N_0 = 0$ and N_t is finite for any t.
(2) The claim sizes are mutually independent and they are also independent of the claim-arrival times.
(3) The claim sizes are identically distributed.
(4) The number or claims in a time interval only depends on the length of the interval, that is

$$P(N_{t_1+\Delta t} - N_{t_1} = k) = P(N_{t_2+\Delta t} - N_{t_2} = k) \text{ for any } t_1, t_2 \ge 0.$$

(5) The number or claims in nonoverlapping intervals are independent. That is, if the intervals $[t_i, t_i']$ are nonoverlapping, then the random variables $N_{t_i'} - N_{t_i}$ are mutually independent.

A model which satisfies these assumptions is called Cramér–Lundberg or *classical collective risk model*. The last two assumptions imply that N_t is a Poisson process with intensity $\beta = E(N_1)$. Hence, the time between the arrival of two consecutive claims is exponentially distributed with parameter β and

$$P(N_{t+h} - N_t = k) = \frac{(\beta h)^k}{k!} e^{-\beta h}.$$

The process $Y_t = \sum_{i=1}^{N_t} U_i$ which corresponds to the total amount of claims paid up to time t is a compound Poisson process. For more details on Poisson and compound Poisson processes, see for instance Varadhan [66].

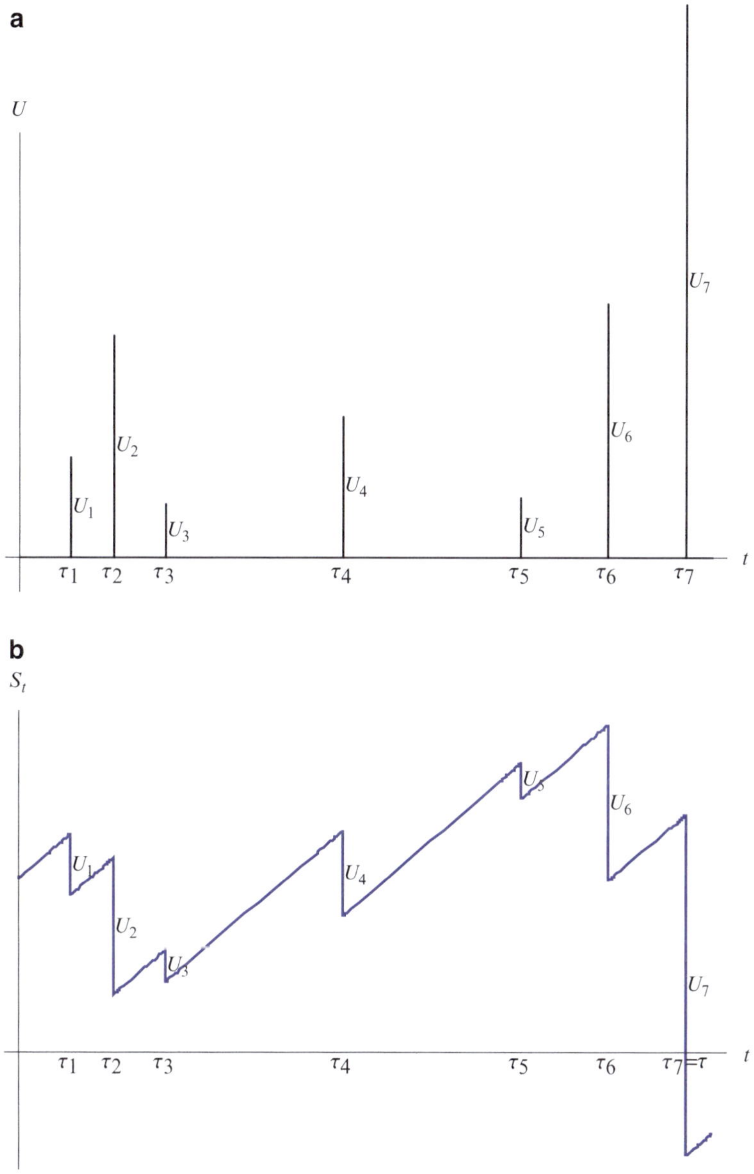

Fig. 1.1 (**a**) A realization of the claim process. (**b**) A realization of the surplus process S_t

The classical collective risk model was introduced by Lundberg [46] in 1903 and developed by Crámer [20] in 1930. This model is completely determined by the premium rate p, the intensity β, and the claim-size distribution function $F(x) = P(U_i \leq x)$. We can describe this model in a rigorous way by defining its filtered probability space $(\Omega, \Sigma, (\mathcal{F}_t)_{t\geq 0}, P)$. We define first the probability space as a product

$$(\Omega, \Sigma, P) = (\Omega_1, \Sigma_1, P_1) \times (\Omega_2, \Sigma_2, P_2). \tag{1.3}$$

The sample space Ω_1 is the set of positive and increasing sequences with infinite limit, Σ_1 is the σ-algebra generated by the sets

$$A_{i,t} = \{(\tau_j)_{j\in\mathbf{N}} \in \Omega_1 \colon \tau_i \leq t\}$$

for $i \in \mathbf{N}$, and $t \geq 0$ and P_1 is the unique probability measure which satisfies

$$P_1(N_t = k) = P_1\Big(\Big(\bigcap_{i=1}^{k} A_{i,t}\Big) \cap A_{k+1,t}^c\Big) = \frac{(\beta t)^k}{k!} e^{-\beta t}.$$

The sample space Ω_2 is the space of sequences

$$\Omega_2 = \{(U_j)_{j\in\mathbf{N}} \in \mathbf{R}_+^{\mathbf{N}}\},$$

where $\mathbf{R}_+ = [0, \infty)$, Σ_2 is the σ-algebra generated by the sets

$$B_{i,a} = \{(U_j)_{j\in\mathbf{N}} \in \Omega_2 \colon U_i \leq a\}$$

for $i \in \mathbf{N}$ and $a \geq 0$ and P_2 is the unique probability measure which satisfies

$$P_2\Big(\Big(\bigcap_{i=1}^{k} B_{i,a_i}\Big)\Big) = \prod_{i=1}^{k} F(a_i) \text{ for } (a_1, .., a_k) \in \mathbf{R}_+^k.$$

Finally, we define the filtration $(\mathcal{F}_t)_{t\geq 0}$ where $\mathcal{F}_t$ is the σ-algebra generated by the random variables τ_i and U_i for all $i \in \mathbf{N}$ such that $\tau_i \leq t$.

The surplus X_t is an adapted càdlàg (left continuous with right limits) stochastic process and the ruin time τ is a stopping time with respect to the filtration $(\mathcal{F}_t)_{t\geq 0}$. Note that the surplus process satisfies *the strong Markov property*. That is, given any stopping time $\overline{\tau}$ with respect to $(\mathcal{F}_t)_{t\geq 0}$ such that $P(\overline{\tau} = \infty) = 0$, the process $X_{\overline{\tau}+t} - X_{\overline{\tau}}$ is independent of $\mathcal{F}_{\overline{\tau}}$ and

$$P(X_{\overline{\tau}+t} - X_{\overline{\tau}} \leq a | \mathcal{F}_{\overline{\tau}}) = P(X_t - X_0 \leq a)$$

for all $a \geq 0$. Roughly speaking, X_t is time homogeneous and memoryless.

We assume that $E(U_i)$ is finite and that the premium rate p is calculated using the *expected value principle* with relative *safety loading* $\eta > 0$, that is

$$p = (1+\eta)E(\sum_{i=1}^{N_1} U_i) = (1+\eta)\beta E(U_i). \tag{1.4}$$

Then we have that $p > \beta E(U_i)$.

Remark 1.1. As a consequence of the strong law of large numbers we have that, given an initial surplus x, $(X_t - x)/t \to p - \beta E(U_i)$ a.s. (see for instance Theorem 2.4 in Durrett [25]) and so we have that the surplus process X_t goes to infinity with probability one.

1.2 Definitions of the Value Functions

One of the central problems associated with the classical collective risk model is to study the ruin probability function ψ. We define

$$\psi(x) = P(\tau < \infty | X_0 = x) \tag{1.5}$$

as the probability that the ruin ever happens in relation to the initial surplus x. This function is an indication of the longterm viability of the business. However, it does not take into account the size of the deficit at the ruin time and also it does not distinguish whether or not the ruin occurs in the near future. The non-ruin or *survival probability function* is defined as

$$\delta(x) = 1 - \psi(x) = P(\tau = \infty | X_0 = x). \tag{1.6}$$

De Finetti [24] proposed in 1957 a more realistic and economically motivated stability criterion: the management of the company should look for maximizing the expectation of the present value of all dividends paid to the shareholders up to ruin time. This criterion involves the problem of finding the optimal dividend strategy.

A *dividend strategy* is a process $\overline{L} = (L_t)_{t\geq 0}$ where L_t is the cumulative dividends the company has paid out until time t; we define the associated *controlled surplus process* $X_t^{\overline{L}}$ as

$$X_t^{\overline{L}} = x + pt - \sum_{i=1}^{N_t} U_i - L_t \tag{1.7}$$

and the corresponding *ruin time* as

$$\tau^{\overline{L}} = \inf\{t \geq 0 : X_t^{\overline{L}} < 0\}.$$

We say that a dividend strategy $\overline{L}$ is *admissible* if it is nondecreasing, càglàd (left continuous with right limits), predictable with respect to the filtration $(\mathcal{F}_t)_{t\geq 0}$, verifies $L_0 = 0$, and

$$L_t \leq X_t = x + pt - \sum_{i=1}^{N_t} U_i \tag{1.8}$$

for any $0 \leq t < \tau^{\overline{L}}$. This last condition says that the company cannot pay immediately an amount of dividends exceeding the current surplus. For technical reasons, we extend the definition of the admissible dividend process as $L_t = L_{\tau^{\overline{L}}}$ for $t \geq \tau^{\overline{L}}$. We denote by Π_x^L the set of all the admissible dividend strategies with initial surplus x. Given any $\overline{L} \in \Pi_x^L$, we have that the controlled surplus process $X_t^{\overline{L}}$ is adapted and the ruin time $\tau^{\overline{L}}$ is a stopping time with respect to the filtration $(\mathcal{F}_t)_{t\geq 0}$.

A jump upwards in the cumulative dividend process L_t at time t_0 means that the company pays immediately the positive sum $L_{t_0^+} - L_{t_0}$ to the shareholders as dividends. Note that the controlled risk process $X_t^{\overline{L}}$ is of finite variation. Also note that $X_{t-}^{\overline{L}} \geq X_t^{\overline{L}} \geq X_{t+}^{\overline{L}}$, where $X_{t-}^{\overline{L}} - X_t^{\overline{L}}$ can only be positive at the arrival of the claims and $X_t^{\overline{L}} - X_{t+}^{\overline{L}}$ is only positive at the discontinuities of L_t and that, by (1.8), the ruin time can only occur at the arrival of a claim.

Given an initial surplus $x \geq 0$ and an admissible dividend strategy $\overline{L} \in \Pi_x^L$, the cumulative expected discounted dividends $V_{\overline{L}}(x)$ is defined as

$$V_{\overline{L}}(x) = E_x\left(\int_0^{\tau^{\overline{L}}} e^{-cs} dL_s\right) := E\left(\int_0^{\tau^{\overline{L}}} e^{-cs} dL_s | X_0 = x\right), \tag{1.9}$$

where $c > 0$ and the integral is interpreted pathwise in a Riemann–Stieltjes sense. The optimal dividend function is defined as

$$V(x) = \sup\{V_{\overline{L}}(x) \text{ with } \overline{L} \in \Pi_x^L\} \text{ for } x \geq 0. \tag{1.10}$$

Remark 1.2. The parameter c in (1.9) is interpreted as the impatience rate of the shareholders. The interest rate has no relation with c because it is assumed that the premium rate, the size of the claims, and the dividend payments are discounted by inflation. See Borch [16] for a discussion on this issue.

1.3 Basic Properties of the Value Functions

In this section we study the regularity and growth at infinity of the value functions of the survival probability and optimal dividend payments problems.

1.3.1 Survival Probability

In the next proposition we prove that the survival probability function δ tends to one as the initial surplus tends to infinity; we also prove that this function is Lipschitz.

Remark 1.3. The survival probability with initial surplus 0 can be calculated explicitly as a special case of the Pollaczek–Khinchine formula (see Chap. 4 of [3])

$$\delta(0) = \frac{p - \beta E(U_i)}{p} = \frac{\eta}{1+\eta} > 0.$$

Note that $\delta(0)$ only depends on the safety loading η and so, it does not depend on the claim-size distribution.

Proposition 1.1. *The survival probability function δ is increasing, Lipschitz, satisfies $0 < \delta(x) < 1$ for $x \in \mathbf{R}_+$ and $\lim_{x\to\infty} \delta(x) = 1$.*

Proof. By definition, δ is nondecreasing. Then, from Remark 1.3, we have that $\delta > 0$. Let us prove that $\delta < 1$ and let us call X_t the process with initial surplus $x_0 \geq 0$. We have that

$$P(X_1 \leq x_0 - 1) = P(\sum_{i=1}^{N_1} U_i \geq p + 1) = P_0 > 0;$$

so we get with a recursive argument that $P(\inf_{t\geq 0} X_t < 0) \geq P_0^{x_0+1}$ and so $\delta(x) \leq 1 - P_0^{x_0+1} < 1$.

We have seen in Remark 1.1 that $X_t \to +\infty$ a.s..Then $X_t - x_0 = pt - \sum_{i=1}^{N_t} U_i$ is lower bounded a.s.. Consider the sets

$$A_n = \left\{ \inf_{t\geq 0} \left(pt - \sum_{l=1}^{N_t} U_i \right) \geq -n \right\}$$

for $n \geq 0$. Since $A_n \subset A_{n+1}$ and the set $\bigcup_{n=0}^{\infty} A_n$ has probability one,

$$\lim_{n\to\infty} \delta(n) = \lim_{n\to\infty} P(A_n) = P(\bigcup_{n=0}^{\infty} A_n) = 1,$$

and so $\lim_{x\to\infty} \delta(x) = 1$.

Let us prove now that δ is increasing. Suppose that $\delta(x_0) = \delta(x_1)$. Let us consider

$$\tau_{x_1} = \inf\{t \geq 0 : X_t = x_1\}.$$

We have, from Remark 1.1, that $P(\tau_{x_1} = \infty) = 0$. Since $\delta(x_0) = \delta(x_1)$ we obtain

$$\delta(x_0) = E_{x_0}(\delta(X_{\tau_{x_1} \wedge \tau})) = \delta(x_1) P(\tau_{x_1} < \tau) = \delta(x_0) P(\tau_{x_1} < \tau)$$

and so, from Remark 1.3, we get that $P(\tau_{x_1} < \tau) = 1$. Note that

$$\tau_{x_1} = \inf\left\{t \geq 0 : pt - \sum_{i=1}^{N_t} U_i = x_1 - x_0\right\},$$

this implies that $\delta(x_1 + (x_1 - x_0)) = \delta(x_1) = \delta(x_0)$. Iterating this procedure we get that $\delta(x_1 + n(x_1 - x_0)) = \delta(x_0)$ for all $n \geq 1$; this is a contradiction since $\delta(x_0) < 1$ and $\lim_{x\to\infty} \delta(x) = 1$.

Let us prove now the Lipschitz property. Consider $0 \leq x_0 \leq x_1$; in the event of no claims the process X_t with initial surplus x_0 reaches x_1 at time $t = (x_1 - x_0)/p$. So we have that $\delta(x_0) \geq \delta(x_1)P(h < \tau_1) = \delta(x_1)e^{-\beta h}$. Then, since δ is bounded by 1, we conclude that

$$0 \leq \delta(x_1) - \delta(x_0) \leq \delta(x_1)(1 - e^{-\beta(x_1 - x_0)/p}) \leq \frac{\beta}{p}(x_1 - x_0).$$

□

Remark 1.4. Since δ is Lipschitz, it is absolutely continuous and differentiable almost everywhere with $0 \leq \delta' \leq \beta/p$ a.e. However, we will see in Chap. 6 that there are examples of claim-size distributions where δ is not differentiable at some points.

1.3.2 Optimal Dividend Payments

In this section we show that the optimal value function V introduced in (1.10) is well defined and describe some of its properties.

Proposition 1.2. *The optimal value function V is well defined and satisfies*

$$x + \frac{p}{c + \beta} \leq V(x) \leq x + \frac{p}{c} \text{ for } x \geq 0.$$

Proof. For any admissible strategy $\overline{L} = (L_t)_{t\geq 0} \in \Pi_x^L$, we have from (1.8) that

$$L_t \leq \varphi(t) := (x + pt)\, I_{\{t \geq 0\}}$$

then, since e^{-ct} is a positive and decreasing function,

$$V_{\overline{L}}(x) \leq E_x\left(\int_0^\infty e^{-ct}\, d\varphi(t)\right) = x + p\int_0^\infty e^{-ct}\, dt = x + \frac{p}{c}.$$

So $V(x) = \sup_{\overline{L}\in\Pi_x^L} V_{\overline{L}}(x)$ is well defined and satisfies the second inequality of the proposition.

Let us prove now the first inequality. Given an initial surplus $x \geq 0$, consider the admissible strategy $\overline{L}_0 \in \Pi_x^L$ which pays x as a lump sum and then pays the incoming premium as dividends until the first claim, which in this strategy means ruin. More precisely, $L_t = x + pt$ for $t \leq \tau_1 = \tau^{\overline{L}_0}$. Then we have

$$V_{\overline{L}_0}(x) = x + pE_x\left(\int_0^{\tau_1} e^{-ct}\, dt\right) = x + \frac{p}{c+\beta},$$

so by (1.10) we get the result. □

Remark 1.5. Note that the above proposition implies in particular that $V(0) \geq p/(c+\beta) > 0$. However it is not clear which is the value of $V(0)$.

Proposition 1.3. *The optimal value function V is increasing and locally Lipschitz in $\mathbf{R}_+$ and satisfies*

$$y - x \leq V(y) - V(x) \leq \beta \frac{V(x)}{p}(y - x)$$

for $y > x \geq 0$.

Proof. Given $\varepsilon > 0$, take an admissible strategy $\overline{L} \in \Pi_x^L$ such that $V_{\overline{L}}(x) \geq V(x) - \varepsilon$. For each $y > x \geq 0$ we define a new strategy $\overline{L}_1 \in \Pi_y^L$ as follows: pay immediately $y - x$ as dividends and then follow the strategy $\overline{L}$. The strategy $\overline{L}_1$ is admissible and we have

$$V(y) \geq V_{\overline{L}_1}(y) = V_{\overline{L}}(x) + (y - x) \geq V(x) - \varepsilon + (y - x),$$

so we obtain the first inequality. Let us prove the second inequality. Given initial surpluses $y > x \geq 0$ and $\varepsilon > 0$, consider an admissible strategy $\overline{L} \in \Pi_y^L$ such that $V_{\overline{L}}(y) \geq V(y) - \varepsilon$. Take now the strategy $\overline{L}_1 \in \Pi_x^L$ which, starting with surplus x, pay no dividends if $X_t^{\overline{L}_1} < y$ and follow strategy $\overline{L}$ after the current surplus reaches y. The strategy $\overline{L}_1$ is admissible. In the event of no claims, the surplus $X_t^{\overline{L}_1}$ reaches y at time $t_0 = (y - x)/p$; then, since the probability of reaching y before the arrival of the first claim is $e^{-\beta t_0}$, we get

$$V(x) \geq V_{\overline{L}_1}(x) \geq V_{\overline{L}}(y)e^{-(c+\beta)t_0} \geq (V(y) - \varepsilon)\, e^{-(\beta+c)(y-x)/p}.$$

Hence we obtain the result. □

Remark 1.6. Since V is Lipschitz on compact sets, it is absolutely continuous with $1 \leq V' \leq (\beta/p)\, V$ a.e. We will prove in Chap. 5 that V is Lipschitz and that there exists $x_0 \geq 0$ such that $V(x) = x - x_0 + V(x_0)$ for $x \geq x_0$.

1.4 HJB Equations

In this section, we find heuristically the first-order integrodifferential equations which satisfy the value function of the stability criteria defined above. To obtain these equations we assume some regularity on the value functions; in the problem of dividend payments we also assume the existence of the optimal dividend strategy. We also need the notion of infinitesimal generator. We will show in Chap. 3 that the value functions could not have enough regularity but still satisfy these equations in a weaker sense.

Definition 1.1. The *infinitesimal generator* $\mathcal{G}$ of a Markov process $\overline{S} = (S_t)_{t\geq 0}$ with $S_0 = x$ is the operator defined on the continuously differentiable functions by

$$\mathcal{G}\left(\overline{S}, f\right)(x) = \lim_{t\to 0} \frac{E_x(f(S_t)) - f(x)}{t}.$$

For more details on infinitesimal generators, see [66].

Let us compute the infinitesimal generator of the process $(X_{t\wedge\tau})_{t\geq 0}$ with initial surplus x. Take any $t > 0$ small enough and consider a continuously differentiable function f in $\mathbf{R}_+$ extended as $f = 0$ for $x < 0$. Let τ_i be the time of the ith claim. Let us define $A_0 = \{\tau_1 > t\}$, $A_1 = \{\tau_1 \leq t, \tau_2 > t\}$ and $A_2 = \{\tau_2 \leq t\}$. Note that $P(A_0) = e^{-\beta t}$, $P(A_1) = (\beta t)\, e^{-\beta t}$ and $P(A_1) = 1 - (1 + \beta t)\, e^{-\beta t} = o(t)$. Then

$$\begin{aligned} E\left(f(X_{t\wedge\tau})\right) &= E\left(f(X_{t\wedge\tau}) I_{A_0}\right) + E\left(f(X_{t\wedge\tau}) I_{A_1}\right) + E\left(f(X_{t\wedge\tau}) I_{A_2}\right) \\ &= e^{-\beta t}\ f(x + pt) + \int_0^t \beta \int_0^{x+ps} f(x + ps - \alpha) dF(\alpha) e^{-\beta s} ds \\ &\quad + o(t) \end{aligned}$$

because f is bounded in $(-\infty, x + pt]$. So we get

$$\mathcal{G}((X_{t\wedge\tau})_{t\geq 0}, f)(x) = p f'(x) - \beta f(x) + \beta \mathcal{I}(f)(x), \tag{1.11}$$

where the integral operator $\mathcal{I}$ is defined by

$$\mathcal{I}(f)(x) = \int_0^x f(x - \alpha) dF(\alpha). \tag{1.12}$$

Remark 1.7. The same argument holds in the case $p < 0$; in this case the infinitesimal generator is

$$\mathcal{G}((X_{t\wedge\tau})_{t\geq 0}, f)(x) = p f'(x) - \beta f(x) + \beta \mathcal{I}(f)(x^-).$$

Remark 1.8. Consider now the process $(X_{t\wedge\tau_1})_{t\geq 0}$ and a continuously differentiable function f in $\mathbf{R}_+$ extended as $f = 0$ for $x < 0$. We get

$$E\left(f(X_{t\wedge\tau_1})\right) = E\left(f(X_{t\wedge\tau_1})I_{\{\tau_1>t\}}\right) + E\left(f(X_{t\wedge\tau_1})I_{\{\tau_1<t\}}\right)$$

$$= e^{-\beta t} f(x+pt) + \beta \int_0^t \mathcal{I}(f)(x+ps)e^{-\beta s}ds$$

and so we have that the infinitesimal generator $\mathcal{G}((X_{t\wedge\tau_1})_{t\geq 0}, f)(x) = \mathcal{G}((X_{t\wedge\tau})_{t\geq 0}, f)(x)$.

1.4.1 Survival Probability

In order to derive formally the first-order integrodifferential equation associated to the survival probability function δ, we need the following elementary result.

Lemma 1.1. *If we extend the definition of the survival probability function δ as $\delta = 0$ in $(-\infty, 0)$, we have that the process $\delta(X_{t\wedge\tau})$ is a martingale.*

Proof. Given an initial surplus $x \geq 0$. By definition of δ, we have that

$$\delta(X_{t\wedge\tau}) = I_{\{X_u\geq 0 \text{ for } u\in[0,t]\}} P(\{X_u \geq 0 \text{ for } u \in [t,\infty)\} \,|\mathcal{F}_t)$$

$$= I_{\{X_u\geq 0 \text{ for } u\in[0,t]\}} E(I_{\{X_u\geq 0 \text{ for } u\in[t,\infty]\}}|\mathcal{F}_t).$$

So, taking $s \leq t$ we have that

$$E(\delta(X_{t\wedge\tau})|\mathcal{F}_s) = E(I_{\{X_u\geq 0 \text{ for } u\in[0,t]\}} E(I_{\{X_u\geq 0 \text{ for } u\in[t,\infty]\}}|\mathcal{F}_t)|\mathcal{F}_s)$$

$$= I_{\{X_u\geq 0 \text{ for } u\in[0,s]\}} E(E\left(I_{\{X_u\geq 0 \text{ for } u\in[s,\infty)\}}|\mathcal{F}_t\right)|\mathcal{F}_s)$$

$$= I_{\{X_u\geq 0 \text{ for } u\in[0,s]\}} E(I_{\{X_u\geq 0 \text{ for } u\in[s,\infty)\}}|\mathcal{F}_s)$$

$$= \delta(X_{s\wedge\tau}).$$

□

Using the previous lemma and the strong Markov property of the surplus process X_t we have that $\delta(x) = E_x\left(\delta(X_{t\wedge\tau})\right)$. Then, assuming that δ is continuously differentiable in $(0,\infty)$, we obtain that δ is a solution of the differential equation

$$\mathcal{G}((X_{t\wedge\tau})_{t\geq 0}, \delta)(x) = 0.$$

By (1.11), the first-order integrodifferential equation associated to the survival probability function δ is

$$\mathcal{L}_0(\delta)(x) = p\delta'(x) - \beta\delta(x) + \beta\mathcal{I}(\delta)(x) = 0. \tag{1.13}$$

If the surplus process is given by any other Markov process $(S_t)_{t\geq 0}$, the associated equation has always the form $\mathcal{G}((S_{t\wedge\tau})_{t\geq 0}, \delta)(x) = 0$ where $\mathcal{G}$ is the infinitesimal operator of the surplus process stopped at the ruin time. For instance, we will show in Sect. 1.5, that in the case that the surplus process is modelled as a Brownian motion with drift, $\mathcal{G}((S_{t\wedge\tau})_{t\geq 0}, \delta)(x) = 0$ is a second-order ordinary differential equation.

1.4.2 Optimal Dividend Payments

In this section we derive formally the HJB equation associated to the optimal dividend payments problem. Again, it turns to be a first-order nonlinear integrodifferential equation, but this time it is fully nonlinear. In order to derive the equation, we introduce the concept of discounted infinitesimal generator and prove the so-called dynamic programming principle (DPP in short).

Definition 1.2. Given a discounted rate $c > 0$, *the discounted infinitesimal generator* $\tilde{\mathcal{G}}$ of a Markov process $\overline{S} = (S_t)_{t\geq 0}$ with $S_0 = x$ is the operator defined on the continuously differentiable functions by

$$\tilde{\mathcal{G}}\left(\overline{S}, f\right)(x) = \lim_{t\to 0} \frac{E_x(e^{-ct} f(S_t)) - f(x)}{t}.$$

Note that

$$\tilde{\mathcal{G}}\left(\overline{S}, f\right)(x) = \mathcal{G}\left(\overline{S}, f\right)(x) - cf(x). \tag{1.14}$$

The DPP plays, in this optimization problem, the role of Lemma 1.1 in the survival probability problem. This is an elementary result which only uses that the optimal value function V defined in (1.10) is continuous and increasing and deals with measurability issues on near-optimal strategies. The result and the proof work for other models for the surplus process as well.

Lemma 1.2. *For any $x \geq 0$ and any stopping time $\overline{\tau}$, we can write*

$$V(x) = \sup_{\overline{L}\in\Pi_x^L} E_x\left(\int_0^{\overline{\tau}\wedge\tau^{\overline{L}}} e^{-cs} dL_s + e^{-c\left(\overline{\tau}\wedge\tau^{\overline{L}}\right)} V(X^{\overline{L}}_{\overline{\tau}\wedge\tau^{\overline{L}}})\right).$$

Proof. We prove this lemma for the case $\overline{\tau}$ equal to a fixed time $T \geq 0$. The general case follows using standard methods; see for instance [68]. We call

$$v(x,T) = \sup_{\overline{L}\in\Pi_x^L} E_x\left(\int_0^{T\wedge\tau^{\overline{L}}} e^{-cs} dL_s + e^{-c\left(T\wedge\tau^{\overline{L}}\right)} V(X^{\overline{L}}_{T\wedge\tau^{\overline{L}}})\right). \tag{1.15}$$

Let us prove first that $V(x) \leq v(x,T)$. Take any admissible strategy $\overline{L} = (L_t) \in \Pi_x^L$; we can write

$$\begin{aligned} V_{\overline{L}}(x) &= E_x\left(\int_0^{T\wedge\tau^{\overline{L}}} e^{-cs}dL_s\right) + e^{-cT}E_x\left(I_{\{\tau^{\overline{L}}>T\}}E(\int_0^{\tau^{\overline{L}}-T} e^{-cs}dL_{s+T}\Big|_{X_T^{\overline{L}}})\right) \\ &\leq E_x\left(\int_0^{T\wedge\tau^{\overline{L}}} e^{-cs}dL_s\right) + e^{-cT}E_x\left(I_{\{\tau^{\overline{L}}>T\}}V(X_T^{\overline{L}})\right) \\ &= E_x\left(\int_0^{T\wedge\tau^{\overline{L}}} e^{-cs}dL_s\right) + E_x\left(e^{-c\left(T\wedge\tau^{\overline{L}}\right)}V(X^{\overline{L}}_{T\wedge\tau^{\overline{L}}})\right) \\ &\leq v(x,T). \end{aligned}$$

From (1.10) we get the result. Let us prove now that $V(x) \geq v(x,T)$. Given any $\varepsilon > 0$, take an admissible strategy $\overline{L} = (L_t) \in \Pi_x^L$ such that

$$E_x\left(\int_0^{T\wedge\tau^{\overline{L}}} e^{-cs}dL_s + e^{-c\left(T\wedge\tau^{\overline{L}}\right)}V(X^{\overline{L}}_{T\wedge\tau^{\overline{L}}})\right) \geq v(x,T) - \varepsilon/2,$$

where $X_t^{\overline{L}}$ is the corresponding controlled risk process. Since V is increasing and continuous in $[0,+\infty)$, we can find an increasing sequence $(x_i)_{i\in\mathbf{N}}$ with $x_1 = 0$ and $\lim_{i\to\infty} x_i = \infty$ such that if $y \in [x_i, x_{i+1})$, then

$$V(y) - V(x_i) < \frac{\varepsilon}{4} \tag{1.16}$$

for $i \geq 0$.

Take admissible strategies $\overline{L_i} = (L_t^i)_{t\geq 0} \in \Pi_{x_i}^L$ such that $V(x_i) - V_{\overline{L_i}}(x_i) < \varepsilon/4$. We define a new strategy $\overline{L_*} = \left(L_t^*\right)_{t\geq 0}$ in the following way:

- If $\tau^{\overline{L}} \leq T$, take $L_t^* = L_t$ for all $t \geq 0$.
- If $\tau^{\overline{L}} > T$, take $L_t^* = L_t$ for $t \in [0,T]$.
- If $\tau^{\overline{L}} > T$ and $X_T^{\overline{L}} \in [x_i, x_{i+1})$, pay immediately $X_T^{\overline{L}} - x_i$ as dividends at time T, that is $L_{T+} - L_T = X_T^{\overline{L}} - x_i$, and then follow strategy $\overline{L_i}$.

By construction, $\overline{L_*}$ is an admissible strategy and if $X_T^{\overline{L}} \in [x_i, x_{i+1})$, we have

$$V_{\overline{L_*}}(X_T^{\overline{L}}) = X_T^{\overline{L}} - x_i + V_{\overline{L_i}}(x_i) \geq V(x_i) - \frac{\varepsilon}{4}. \tag{1.17}$$

Using (1.15)–(1.17), we obtain

$$v(x,T)-V_{\overline{L}_*}(x) \leq E_x\left(\int_0^{T\wedge\tau^{\overline{L}}} e^{-cs}dL_s+e^{-c(T\wedge\tau^{\overline{L}})}V(X^{\overline{L}}_{T\wedge\tau^{\overline{L}}})\right)-V_{\overline{L}_*}(x)+\frac{\varepsilon}{2}<\varepsilon,$$

and so we get the result. □

Our objective is to find the first-order integrodifferential equation which satisfies the value function V. In order to do this we show that, if V is continuously differentiable, then it satisfies an integrodifferential inequality; the equality is what is called the HJB equation. In Chap. 3, we show that V is indeed a viscosity solution of this equation.

Assume that V is continuously differentiable at x. Given any $l \geq 0$, let us consider the admissible strategy $\overline{L}$ which pays dividends at constant rate l. Let us call the corresponding controlled surplus process $X_t^{\overline{L}} = X_t - lt$ and the corresponding ruin time τ. The surplus process $X^{\overline{L}}_{\tau\wedge t}$ stopped at the ruin time is a Markov process and $X^{\overline{L}}_{t\wedge\tau}$ is the same process as $X_{t\wedge\tau}$ but with drift $(p-l)$, so from (1.11), and Remarks 1.7 and 1.8, we can write

$$\tilde{\mathcal{G}}\left(\left(X^{\overline{L}}_{t\wedge\tau_1}\right)_{t\geq 0}, V\right)(x) = \begin{cases} (p-l)\,V'(x)-(\beta+c)\,V(x)+\beta\mathcal{I}(V)(x) & \text{if } l \leq p \\ (p-l)\,V'(x)-(\beta+c)\,V(x)+\beta\mathcal{I}(V)(x^-) & \text{if } l > p. \end{cases} \tag{1.18}$$

Take any $t > 0$ such that $t < x/(l-p)$ in the case of $p < l$; using Lemma 1.2 we get

$$V(x) \geq E_x(\int_0^{\tau_1\wedge t} e^{-c\,s}\,l\,ds) + E_x\left(e^{-c\,(\tau_1\wedge t)}V(X^{\overline{L}}_{t\wedge\tau_1})\right).$$

Then

$$0 \geq \lim_{t\to 0^+}\left(\frac{l E_x(\int_0^{\tau_1\wedge t} e^{-c\,s}\,ds)}{t}+\frac{E_x\left(e^{-c\,(\tau_1\wedge t)}V(X^{\overline{L}}_{t\wedge\tau_1})\right)-V(x)}{t}\right)$$

$$= l+\tilde{\mathcal{G}}\left(\left(X^{\overline{L}}_{t\wedge\tau_1}\right)_{t\geq 0}, V\right)(x)$$

because

$$\begin{aligned} E_x(\textstyle\int_0^{\tau_1\wedge t} e^{-c\,s}\,ds) &= E_x(I_{\tau_1<t}\textstyle\int_0^{\tau_1} e^{-c\,s}\,ds)+E_x(I_{\tau_1\geq t}\textstyle\int_0^t e^{-c\,s}\,ds) \\ &= \textstyle\int_0^t \beta e^{-\beta u}\left(\int_0^u e^{-c\,s}\,ds\right)du+e^{-\beta t}\int_0^t e^{-c\,s}ds \\ &= \frac{1-e^{-(c+\beta)\,t}}{\beta+c}. \end{aligned}$$

So we obtain the inequality

$$\sup_{l \geq 0} \left\{ l + \tilde{\mathcal{G}} \left(\left(X_{t \wedge \tau}^{\overline{L}} \right)_{t \geq 0}, V \right) (x) \right\} \leq 0. \tag{1.19}$$

The HJB equation of this optimization problem is

$$\sup_{l \geq 0} \left\{ l + \tilde{\mathcal{G}} \left(\left(X_{t \wedge \tau}^{\overline{L}} \right)_{t \geq 0}, V \right) (x) \right\} = 0. \tag{1.20}$$

Using the formula for the discounted infinitesimal generator (1.18) we obtain

$$l + \tilde{\mathcal{G}} \left(\left(X_{t \wedge \tau}^{\overline{L}} \right)_{t \geq 0}, V \right) (x) = H(l),$$

where

$$H(l) = \begin{cases} l\,(1 - V'(x)) + pV'(x) - (c + \beta)V(x) + \beta \mathcal{I}(V)(x) & \text{if } l \leq p \\ l\,(1 - V'(x)) + pV'(x) - (c + \beta)V(x) + \beta \mathcal{I}(V)(x^-) & \text{if } l > p. \end{cases}$$

Note that $V'(x) \geq 1$; because if this were not the case, $H(l)$ would be positive for l large enough. The function H is decreasing and has a downward jump at $l = p$ in the case that F is not continuous at x; so the maximum of $H(l)$ is attained at $l = 0$. Therefore, the HJB equation of this problem can be rewritten as

$$\max\{1 - V'(x), \tilde{\mathcal{L}}_0(V)(x)\} = 0, \tag{1.21}$$

where

$$\tilde{\mathcal{L}}_0(V)(x) = pV'(x) - (c + \beta)V(x) + \beta \mathcal{I}(V)(x). \tag{1.22}$$

Note that $\tilde{\mathcal{L}}_0(V)(x)$ is the discounted infinitesimal generator of the uncontrolled surplus process applied to V, that is $\tilde{\mathcal{L}}_0(V)(x) = \tilde{\mathcal{G}}\left((X_{t \wedge \tau})_{t \geq 0}, V\right)(x)$.

1.5 A Limit Case: Diffusion Approximation

In the case of big portfolios, when the claims have small sizes and arrive with high frequency, the classical risk surplus process $X_t - X_0$ can be approximated by a Brownian motion with drift. This approximation, due to Iglehart [36], is obtained using a functional central limit theorem, rescaling in a special way the intensity of the claim-arrival time, the claim-size distribution, and the safety loading.

Given β and η positives and a distribution function F on $\mathbf{R}_+$ with finite mean μ and second moment γ^2, let us consider the sequence of rescaled risk surplus processes:

$$R_t^n = (1+\eta_n)\beta_n E(U_i^n)t - \sum_{i=1}^{N_t^{\beta_n}} U_i^n = \eta\beta\mu t + M_t^n$$

with claim intensity $\beta_n = \beta n$, claims $U_i^n = U_i/\sqrt{n}$ with distribution function $F_n(x) = F(\sqrt{n}x)$, and safety loading $\eta_n = \eta/\sqrt{n}$. Then we can write $R_t^n = \eta\beta\mu t + M_t^n$, where

$$M_t^n = \frac{\beta\mu n t - \sum_{i=1}^{N_{nt}} U_i}{\sqrt{n}}$$

is a martingale with zero mean and variance $\beta\gamma^2 t$. Iglehart proved that the processes M_t^n converge weakly in the space of càdlàg processes with the Skorohod topology to a Brownian motion with variance rate $\beta\gamma^2$. That is

$$R_t^n \Longrightarrow \overline{R}_t = \nu t + \sigma W_t$$

where W_t is the standard Brownian motion, $\nu = \eta\beta\mu$, and $\overline{\sigma} = \sqrt{\beta}\gamma$. Here the symbol $\Longrightarrow$ denotes weak convergence. Note that with this rescaling procedure, not only the claim-arrival intensity β_n goes to infinity and the claim-size mean $\mu_n = E(U_i^n)$ goes to zero as n goes to infinity, we also have that the safety loading η_n goes to 0. Also note that the premium rate $p_n = (1+\eta_n)\beta_n\mu_n$ can be think as a sum of two terms:

- $\eta_n\beta_n\mu_n = \nu$, that is the drift of the limit Brownian motion
- $\beta_n\mu_n = \beta\mu\sqrt{n}$, which compensates the cumulative claims to make M_1^n a martingale.

In the limit model for the surplus, namely

$$\overline{X}_t = x + \nu t + \overline{\sigma} W_t, \tag{1.23}$$

the trajectories are continuous and the ruin time τ coincides with the first time that the process $\overline{X}_t$ reaches zero.

Let us derive the formula of the infinitesimal generator for the process $\overline{X}_{t\wedge\tau}$. Consider f a twice continuously differentiable function , with bounded derivative; then by Itô's formula we get

$$f(\overline{X}_{t\wedge\tau}) - f(x) = \int_0^{t\wedge\tau} f'(\overline{X}_t)\,(\nu dt + \overline{\sigma} d W_t) + \int_0^{t\wedge\tau} f''(\overline{X}_t)\frac{\overline{\sigma}^2}{2} dt.$$

Taking expectation and using that W_t is a martingale, we obtain

$$E_x(f(\overline{X}_{t\wedge\tau})) - f(x) = E_x\left(\int_0^{h\wedge\tau}\left(\frac{\overline{\sigma}^2}{2}f''(\overline{X}_t) + \nu f'(\overline{X}_t)\right)dt\right).$$

Dividing by h and taking $h \to 0^+$, we get

$$\mathcal{G}((\overline{X}_{t\wedge\tau})_{t\geq 0}, f) := \mathcal{L}_D(f)(x) := \frac{\overline{\sigma}^2}{2}f''(x) + \nu f'(x) \tag{1.24}$$

for all $x \geq 0$.

1.5.1 *Survival Probability*

In this setting, the survival probability function δ can be obtained explicitly using standard Brownian motion computations. However, we can derive the equation associated to δ as in Sect. 1.4.1. Assume that δ is twice continuously differentiable in $(0,\infty)$ with bounded derivative; since $\delta(\overline{X}_{t\wedge\tau})$ is a martingale we have by Definition 1.1 that $\mathcal{G}((\overline{X}_{t\wedge\tau})_{t\geq 0}, \delta) = 0$ and so from (1.24) we get formally

$$\mathcal{L}_D(\delta)(x) = 0 \tag{1.25}$$

for all $x \geq 0$. We can see immediately that $\delta(0) = 0$ and, with a similar proof of Proposition 1.1, we obtain that $\lim_{x\to\infty}\delta(x) = 1$. Hence, these are the natural boundary conditions for this problem. The function $1 - e^{-2\nu x/\overline{\sigma}^2}$ is the unique solution of (1.25) with these boundary conditions. The survival probability function δ in this setting turns to be twice continuously differentiable so $\delta(x) = 1 - e^{-2\nu x/\overline{\sigma}^2}$.

1.5.2 *Optimal Dividend Payments*

Let us derive now the HJB equation for the optimal dividend payments problem in the limit diffusion model. Given a *dividend strategy* $\overline{L} = (L_t)_{t\geq 0}$, we define *the controlled surplus process* $\overline{X}_t^{\overline{L}}$ as

$$\overline{X}_t^{\overline{L}} = \overline{X}_t - L_t. \tag{1.26}$$

and the corresponding *ruin time* as

$$\tau^{\overline{L}} = \inf\{t \geq 0 : X_t^{\overline{L}} < 0\}.$$

A dividend strategy $\overline{L}$ is *admissible* if it is nondecreasing, càglàd, predictable with respect to the filtration $\left(\mathcal{F}_t\right)_{t\geq 0}$ generated by the Brownian motion W_t, and satisfies $L_0 = 0$ and $L_t \leq \overline{X}_t$ for any $0 \leq t < \tau^{\overline{L}}$. For any $\overline{L} \in \Pi_x^L$, note that $X_{t-}^{\overline{L}} = X_t^{\overline{L}} \geq X_{t+}^{\overline{L}}$, where $X_t^{\overline{L}} - X_{t+}^{\overline{L}}$ is positive at the discontinuities of L_t.

From (1.14) and (1.25), we get

$$\tilde{\mathcal{G}}(\left(\overline{X}_{t\wedge\tau}\right)_{t\geq 0}, f)(x) = \tilde{\mathcal{L}}_D(f)(x) := \frac{\overline{\sigma}^2}{2} f''(x) + \nu f'(x) - c f(x). \qquad (1.27)$$

The DPP stated in Lemma 1.2 holds for the limit diffusion process $\overline{X}_{t\wedge\tau}$. We assume that V is twice continuously differentiable at x with bounded derivative. Given any $l \geq 0$, we consider the admissible strategy $\overline{L_0} = (tl)_{t\geq 0} \in \Pi_x^L$. We can obtain, as in (1.20), the HJB equation

$$\sup_{l\geq 0}\left\{ l + \tilde{\mathcal{G}}\left((\overline{X}_{t\wedge\tau}^{L_0})_{t\geq 0}, V \right)(x) \right\} = 0.$$

We have that $l + \tilde{\mathcal{G}}\left((\overline{X}_{t\wedge\tau}^{L_0})_{t\geq 0}, V \right)(x) = \overline{H}(l) \leq 0$, where

$$\overline{H}(l) = l\left(1 - V'(x)\right) + \frac{\overline{\sigma}^2}{2} V''(x) + \nu V'(x) - c V(x).$$

Since these inequalities hold for all $l \geq 0$, then $V'(x) \geq 1$ and the maximum of $\overline{H}(l)$ is attained at $l = 0$. So we obtain that the HJB equation of the optimal dividend payments problem for the limit diffusion case could also be written as

$$\max\{1 - V'(x), \tilde{\mathcal{L}}_D(V)(x)\} = 0. \qquad (1.28)$$

Here, unlike in the classical risk model for the surplus process, there is a natural boundary condition at zero $V(0) = 0$ because the ruin time is zero regardless of the dividend strategy.

1.6 Discussion on the Characterization of Value Functions

In Sects. 1.4 and 1.5, we have derived formally the equations associated to the survival probability and the optimal dividend payments problems in both the classical risk model and the limit diffusion one. The best scenario possible would be for the value function to be the unique solution of the corresponding equation under the natural boundary conditions. Besides, in the optimal dividend payments problems, one would expect to obtain the optimal strategy from the optimal value function.

In the limit diffusion model presented in Sect. 1.5, the associated equations are simpler because they do not involve integrals and its solutions can be obtained explicitly. Moreover, using standard martingale techniques for Brownian motion, a *verification theorem* can be proved: a solution of the associated equation satisfying the natural boundary conditions has to be the optimal value function. Moreover, the optimal dividend strategy exists and it has a very simple structure. So, this is the best scenario possible. This problem was solved in Shreve et al. [58] and in Asmussen and Taksar [4].

Things are more complicated in the classical setting. The first issue is whether the value function is smooth enough to be a solution of the associated equation. In the case of the survival probability function, this would be the case when the claim-size distribution function F is continuous. However, it is not hard to see that a differentiable function cannot be a solution of (1.13) at the points where F is not continuous and so δ cannot be a smooth function. The situation is even worse in the problem of optimal dividend payments: in Chap. 6 we will show an example where the claim-size distribution F is continuous (in fact it has a bounded density) and nevertheless the value function V is not differentiable. Therefore, the standard framework is not broad enough to include the solutions of our problems. To overcome this difficulty, we consider the notion of viscosity solution. The main characteristic of this notion is to replace the conventional derivatives by super- and sub-differentials, so it allows non-smooth solutions of the HJB equations. This does not mean that a viscosity solution is just a solution almost everywhere; in general if the equation (together with the boundary conditions) has a unique classical solution, this classical solution is also the unique viscosity solution. In Chap. 3, we will define viscosity solutions and show that the value functions are indeed viscosity solutions of the corresponding equation.

The second issue is whether the HJB equations associated to the problems have unique viscosity solutions under the natural boundary conditions. We have seen in Sect. 1.3 that the natural boundary condition for the survival probability function would be $\lim_{x\to\infty} \delta(x) = 1$ and in the case of optimal dividend payments would be $\lim_{x\to\infty} (V(x) - x) \leq p/c$. The uniqueness of viscosity solution gives a *characterization result* since any viscosity solution of the associated equation that satisfies the natural boundary condition at infinity should be the value function. It also gives what is called a *verification theorem*, namely if a smooth function is a solution of the equation, this function is the value function. We will show in Chap. 4 that there exists a unique viscosity solution for the equation associated to the survival probability problem. However, in the optimal dividend payments one, there are infinitely many viscosity solutions of the HJB equation (1.21) that satisfy the growth condition at infinity. Therefore, we need another way to characterize the value function V among all the viscosity solutions: we will prove in Chap. 4 that the value function can be characterized as the smallest viscosity solution of the HJB equation.

Finally, in the optimal dividend payments problem, the remaining issue is the existence of optimal strategies and their structure. In the limit diffusion case the optimal strategies exist and have a very simple structure: they are barrier strategies.

A *barrier dividend strategy* with level $a \geq 0$ pays immediately as dividends all the surplus above a and then pays as dividends all the incoming premiums which make the surplus to surpass the level a. In particular, if the current surplus is bellow a the barrier, strategy pays no dividends. Again, things are more complicated in the classical risk model. We will prove in Chap. 5 that, for general claim-size distributions, there exist optimal strategies and that they are in fact band strategies (not necessarily barrier).

1.7 Comments and References

We choose to construct the filtered probability space $(\Omega, \Sigma, (\mathcal{F}_t)_{t\geq 0}, P)$ using the compound Poisson process as in Sect. 2.2 of [66]; another possibility would be to view the stochastic processes as probability measures on the Polish space of càdlàg functions; see for instance Appendix A of [66].

The optimal dividend problem in the classical risk model was first solved by Gerber [29] via a discretization of the time parameter and the claim distribution function and then it was reconsidered by Azcue and Muler [9] using stochastic control theory; in this work we take the second approach.

In the limit diffusion approximation, the optimal dividend problem was address by Shreve et al. [58], Jeanblank-Picqué and Shiryaev [38], and Asmussen and Taksar [4]. Both the classical and the diffusion risk models are special cases of general spectrally negative Lévy risk processes. The optimal dividend problem in this more general setting was studied by Avram et al. [7], Loeffen [43], and Kyprianou et al. [41].

We have addressed in this chapter the bare problems. Let us mention some related problems that are beyond the scope of this work although the tools that we will present in Chaps. 3–5 can be used to study them.

The optimal dividend problem with a ceiling on the dividend rates was studied in the diffusion risk model by Asmussen and Taksar [4], in the classical risk model by Gerber and Shiu [30] for exponential claim-size distributions and by Azcue and Muler [12] for general claim-size distributions.

There are several possibilities to extend the definition of the optimal value function in the dividend payments problem. One is to combine the two stability criteria maximizing the expected cumulative discounted dividends but considering a penalty payment at the moment of ruin (which is an increasing function of the size of the shortfall at ruin) or continuous payoffs until ruin; we can mention the works of Dickson and Waters [23], Gerber et al. [31], Albrecher and Thonhauser [1], Cai et al. [19], and Loeffen and Renaud [45], among others. Another possibility is to avoid ruin with capital injection; see for instance Avram et al. [7] and Kulenko and Schmidli [40]. Finally, the introduction of transaction costs leads to impulse controls; see for instance Loeffen [44], Tonhauser and Albrecher [63], and Avram et al. [8].

For an exhaustive study on ruin probability, including the cases of classical, diffusion, and general Lévy risk processes, see the book by Asmussen and Albrecher [3].

Finally, let us mention the surveys of Avanzi [6] and Albrecher and Thonhauser [2] and the book of Schmidli [57] which covers most of the topics of this work from a different perspective.

Chapter 2
Reinsurance and Investment

In this chapter we present the two main ways to control the insurance risk process: reinsurance and investment. We focus on the classical risk model.

2.1 Reinsurance in the Classical Risk Model

An insurance company can share the risk by a reinsurance contract. We only consider the case in which this contract reduces the impact of each one of the claims, that is, by paying to the reinsurance company some part of the premium; this company covers some predetermined part of the claim. A reinsurance contract has two elements:

- A Borel measurable function $R : \mathbf{R}_+ \to \mathbf{R}_+$ (called *retained loss function*) that satisfies $0 \leq R(\alpha) \leq \alpha$, where $R(\alpha)$ is the part of the claim paid by the insurance company when the size of the claim is α (the reinsurance company covers $\alpha - R(\alpha)$) .
- The premium rate q_R paid to the reinsurance company. So the premium rate left to the insurance company is $p_R = p - q_R$.

The part of the claim paid by the insurance company is the random variable $R(U)$ where U is the claim size. We define

$$F_R(x) = P(R(U) \leq x). \tag{2.1}$$

The two more common examples of reinsurance contracts are *proportional* reinsurance and *excess-of-loss* reinsurance. In the first case the reinsurance company covers a fixed ratio of the claim and therefore the retained loss function is $R(\alpha) = b\alpha$ for some *retained proportion* $b \in [0, 1]$; here $F_R(x) = F(x/b)$ if $b > 0$ and $F_R(x) = 1$ if $b = 0$. In the second case a *retention level* $a \in [0, \infty]$ is fixed in such a way

P. Azcue and N. Muler, *Stochastic Optimization in Insurance: A Dynamic Programming Approach*, SpringerBriefs in Quantitative Finance, DOI 10.1007/978-1-4939-0995-7_2,

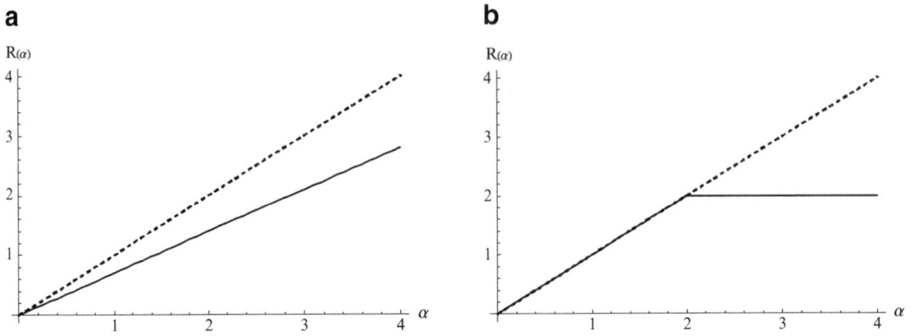

Fig. 2.1 (**a**) Proportional reinsurance. (**b**) Excess-of-loss reinsurance

that, paying to the reinsurance company some part of the premium, the reinsurance company covers the amount of the claim exceeding a; in this case the retained loss function is $R(\alpha) = \min\{\alpha, a\}$ and

$$F_R(x) = F(x)I_{\{x<a\}} + I_{\{x\geq a\}}.$$

We show the graphs of the retained loss function of proportional and excess-of-loss reinsurance contracts corresponding to $b = 0.7$ and $a = 2$ in Fig. 2.1a, b, respectively (the identity function is shown in dotted line).

We assume that the premium rate of the reinsurance company is calculated using the expected value principle with a relative safety loading $\eta_1 \geq \eta > 0$; we obtain that

$$q_R = (1+\eta_1)\beta E(U_i - R(U_i))$$

and so

$$\begin{aligned} p_R &= (1+\eta)\beta E(U_i) - (1+\eta_1)\beta E(U_i - R(U_i)) \\ &= (1+\eta_1)\beta E(R(U_i)) - (\eta_1 - \eta)\beta E(U_i). \end{aligned} \tag{2.2}$$

The case $\eta_1 = \eta$ is called cheap reinsurance. There are other criteria for computing the premium rate of the reinsurance company; see for instance Teugels [62]. We assume that the premium rate left to the insurance company is $p_R = p - q_R > 0$.

Definition 2.1. Let us call $\mathcal{R}_A$ the family of all the retained loss functions with positive p_R, $\mathcal{R}_P \subset \mathcal{R}_A$ the subfamily of proportional retained loss functions, and $\mathcal{R}_{XL} \subset \mathcal{R}_A$ the subfamily of the excess-of-loss retained loss functions. We denote by $\mathcal{R}_F$ any finite subfamily of retained functions in $\mathcal{R}_A$.

Given any subfamily $\mathcal{R} \subset \mathcal{R}_A$ of retained loss functions, we assume that the manager of the insurance company can choose at any time a reinsurance contract within the family $\mathcal{R}$ and that the premium rate of the reinsurance company is calculated using the expected value principle with relative safety loading $\eta_1 > \eta$.

Consider the filtered probability space $(\Omega, \Sigma, (\mathcal{F}_t)_{t\geq 0}, P)$ introduced in (1.3). A reinsurance control strategy is a collection $\overline{R} = (R_t)_{t\geq 0}$ of functions $R_t : \Omega \to \mathcal{R}$ for any $t \geq 0$. We say that a reinsurance control strategy is *admissible* if the function $(\omega, \alpha, t) \to R_t(\omega)(\alpha)$ is $(\Sigma \times \text{Borel} \times \text{Borel})$ measurable and the function $\omega \to R_t(\omega)(\alpha)$ is $\mathcal{F}_{t-}$ measurable for every $t \geq 0$ and $\alpha \geq 0$. Note that this definition means that the process $(R_t(\cdot)(\alpha))_{t\geq 0}$ is predictable for any α. We denote by Π_x^R the set of all the admissible control strategies with initial surplus $x \geq 0$. Note that, for any reinsurance admissible control strategies $\overline{R} \in \Pi_x^R$, the premium process $(p_{R_t})_{t\geq 0}$ is Borel measurable.

Given an admissible control strategy $\overline{R}$, the *controlled risk process* $X_t^{\overline{R}}$ is given by

$$X_t^{\overline{R}} = x + \int_0^t p_{R_s}\, ds - \sum_{i=1}^{N_t} R_{\tau_i}(U_i), \tag{2.3}$$

where τ_i is the time of occurrence of the ith claim. We define the corresponding *ruin time* $\tau^{\overline{R}}$ of the company as

$$\tau^{\overline{R}} = \inf\left\{t \geq 0 : X_t^{\overline{R}} < 0\right\}. \tag{2.4}$$

An important class of reinsurance admissible strategies is the one where the decision of the reinsurance contract depends only on the current surplus. The idea is the following: consider a retained loss function $\rho^y \in \mathcal{R}$ for each $y \geq 0$, and define for any initial surplus $x \geq 0$, the process $(X_t)_{t\geq 0}$ obtained by taking ρ^y as retained loss function when the current surplus is y; the process X_t should satisfy

$$X_t = x + \int_0^t p_{\rho^{X_{s-}}}\, ds - \sum_{i=1}^{N_t} \rho^{X_{\tau_i^-}}(U_i) \tag{2.5}$$

and it should be the controlled reinsurance process associated with the reinsurance strategy $\left(\rho^{X_{t-}}\right)_{t\geq 0} \in \Pi_x^R$.

We define a stationary reinsurance control as a choice of a retained loss function for each surplus; with the suitable measurability conditions, we obtain that the strategy $\left(\rho^{X_{t-}}\right)_{t\geq 0}$ is admissible and therefore it belongs to Π_x^R. More precisely:

Definition 2.2. A *stationary reinsurance control in* $\mathcal{R}$ is a Borel measurable function $\rho : \mathbf{R}_+ \times \mathbf{R}_+ \to \mathbf{R}$ such that $\rho(x, \cdot) = \rho^x \in \mathcal{R}$ for all $x \geq 0$ and $1/p_{\rho^x}$ is locally integrable.

Proposition 2.1. *Given any stationary reinsurance control ρ and any initial surplus $x \in \mathbf{R}_+$, there exists a unique solution $(X_t)_{t\geq 0}$ of the stochastic integral equation (2.5). Moreover, if we define $R_t = \rho^{X_{t}-}$, then the strategy $(R_t)_{t\geq 0} \in \Pi_x^R$ and its associated surplus process $\left(X_t^R\right)_{t\geq 0}$ coincides with $(X_t)_{t\geq 0}$.*

Proof. In order to see that there exists a unique solution of (2.5), it is enough to show existence a uniqueness for a fix $\omega \in \Omega$ and for t between two claims, that is,

$$X_t = x + \int_0^t p_{\rho^{X_{s^-}}}\, ds. \tag{2.6}$$

Let us define the function

$$G(x) = \int_0^x \frac{1}{p_{\rho^y}} dy;$$

since $p_{\rho^y} \in (0, p]$, the function G is Lipschitz and increasing. The unique solution of (2.6) can be written as

$$X_t = G^{-1}(G(x) + t).$$

On the other hand, since ρ is Borel measurable and the process X_{t^-} is $\mathcal{F}_{t^-}$-measurable, we have that the strategy $\left(\rho^{X_{t^-}}\right)_{t \geq 0}$ is admissible. □

Given an initial surplus $x \geq 0$ and any fixed retained function $R \in \mathcal{R}$, we consider the constant admissible strategy $\overline{R} = (R)_{t \geq 0}$, the corresponding controlled surplus process $X_t^{\overline{R}}$, and the ruin time $\tau^{\overline{R}}$. The process $X_{t \wedge \tau^{\overline{R}}}^{\overline{R}}$ is Markov, by (1.11), its infinitesimal generator is $\mathcal{G}(\left(X_{t \wedge \tau^{\overline{R}}}^{\overline{R}}\right)_{t \geq 0}, f)(x) = \mathcal{L}_R(f)(x)$ where

$$\mathcal{L}_R(f)(x) = p_R f'(x) - \beta f(x) + \beta \mathcal{I}_R(f)(x), \tag{2.7}$$

and

$$\mathcal{I}_R(f)(x) = \textstyle\int_0^\infty f(x - R(\alpha)) dF(\alpha) = \int_0^x f(x - \alpha) dF_R(\alpha) \tag{2.8}$$

(here f is a continuously differentiable function in $\mathbf{R}_+$ extended as $f = 0$ for $x < 0$). This integral is interpreted in the Lebesgue–Stieltjes sense; in the case that the integral exists in the Riemann–Stieltjes sense, both notions agree. In the case that R is either continuous or it has finitely many discontinuities which do not coincide with the discontinuities of F, this integral exists in the Riemann–Stieltjes sense. See Sect. 12.3 in Royden [52].

2.1.1 *Survival Probability and Reinsurance*

We assume in this section that within the family $\mathcal{R}$ there exists at least one retained function $\hat{R} \in \mathcal{R}$ that satisfies the *net profit condition*, that is $p_{\hat{R}} > \beta E(\hat{R}(U_i))$.

Given $x \geq 0$, any admissible control strategy $\overline{R} \in \Pi_x^R$ and its controlled risk process $X_t^{\overline{R}}$, we define the corresponding survival probability as

$$\delta^{\overline{R}}(x) = P(\tau^{\overline{R}} = \infty | X_0^{\overline{R}} = x),$$

where $\tau^{\overline{R}}$ is introduced in (2.4). The *optimal survival probability function with reinsurance* is defined as

$$\delta(x) = \sup_{\overline{R} \in \Pi_x^R} \delta^{\overline{R}}(x). \tag{2.9}$$

We have the following property with respect to the behavior of the surplus $X_t^{\overline{R}}$ at infinity, the proof is similar to the one given in Lemma 2.9 of [57].

Proposition 2.2. *Take any admissible strategy $\overline{R} \in \Pi_x^R$; with probability one, either ruin occurs in finite time or $X_t^{\overline{R}}$ diverges to infinity as t goes to infinity.*

Proof. Suppose that the initial surplus is $x \geq 0$. By (2.2), we have that

$$E(R(U)) \geq \varsigma := \frac{(\eta_1 - \eta)\beta\,(E(U))}{2(1+\eta_1)\beta} > 0$$

for all $R \in \mathcal{R}$. We show now that there exists $\gamma > 0$ small enough such that $P(R(U) \geq \gamma) \geq \gamma > 0$ for all $R \in \mathcal{R}$. Suppose that this is not the case, then for each $n \geq 1$ there exists $R_n \in \mathcal{R}$ such that $P(R_n(U) \geq 1/n) \leq 1/n$. Since $E(U)$ is finite and

$$E(R_n(U)) \leq \int_0^\infty \tfrac{1}{n} I_{\{R_n < \frac{1}{n}\}} dF(\alpha) + \int_0^\infty \alpha I_{\{R_n \geq \frac{1}{n}\}} dF(\alpha),$$

we have

$$0 < \varsigma \leq \limsup_{n\to\infty} E(R_n(U)) \leq \lim_{n\to\infty} \frac{1}{n} + \limsup_{n\to\infty} \int_0^\infty \alpha I_{\{R_n \geq \frac{1}{n}\}} dF(\alpha) = 0$$

and this is a contradiction.

Fix $a > 0$, we define recursively the sequence $(t_k)_{k\in\mathbf{N}}$ in the following way

$$t_1 = \inf\{t \geq 0 : X_t^{\overline{R}} < a\}$$

and

$$t_{k+1} = \inf\{t \geq 0 : t \geq t_k + 1 \text{ and } X_t^{\overline{R}} < a\}.$$

In the case that $\liminf_{t\to\infty} X_t^{\overline{R}} < a$, all the $t_k^{\prime}s$ are finite. In the case that $X_t^{\overline{R}} \geq a$ for all $t \geq 0$, we put $t_n = \infty$ for $n \geq 1$, and in the case that $X_t^{\overline{R}} \geq a$ for all $t \geq t_k + 1$, we put $t_n = \infty$ for $n \geq k + 1$. Let us consider the case $t_k < \infty$; we define Σ_k as the σ-algebra generated by $\left(X_{t\wedge t_k}^{\overline{R}}\right)_{t\geq 0}$. Then, if $\omega \in \Omega$ satisfies $t_k(\omega) < \infty$, we have for any $\varepsilon > 0$ that

$$
\begin{aligned}
& E(I_{\{t_k<\infty \text{ and } X^{\overline{R}}_{t_k+1}\leq a-\varepsilon\}}|\Sigma_k)(\omega) \\
& \geq P(X^{\overline{R}}_{t_k+1} \leq a-\varepsilon | X^{\overline{R}}_{t_k} = a) \\
& = P(\textstyle\sum_{\tau_i\in(t_k,t_k+1]} R_{\tau_i}(U_i) \geq \int_{t_k}^{t_k+1} p_{R_s} ds + \varepsilon) \geq P_0 > 0,
\end{aligned}
$$

because $P(R_{\tau_i}(U_i) \geq \gamma) \geq \gamma > 0$, R_t is predictable, U_i are i.i.d, and N_t is a Poisson process independent on both the arrival claim times and the size of the claims. Hence,

$$
E(I_{\{t_k<\infty \text{ and } X^{\overline{R}}_{t_k+1}\leq a-\varepsilon\}} - P_0 I_{\{t_k<\infty\}}|\Sigma_k) \geq 0. \tag{2.10}
$$

Let us define

$$
W_k = I_{\{t_k<\infty \text{ and } X^{\overline{R}}_{t_k+1}\leq a-\varepsilon\}},\ Z_k = P_0 I_{\{t_k<\infty\}} \text{ and } A_n = \sum_{k=1}^{n}(W_k - Z_k).
$$

Using that $A_1, A_2, \ldots, A_{n-1}$ are Σ_n-measurable and (2.10), we have that A_n is a submartingale, and so from Lemma 2.1 of Niemiro and Pokarowski [48], we obtain that

$$
P\left(\textstyle\sum_{k=1}^{\infty} Z_k = \infty, \sum_{k=1}^{\infty} W_k < \infty\right) = 0.
$$

We conclude that if $\liminf_{t\to\infty} X_t^{\overline{R}} < a$, then $\liminf_{t\to\infty} X_t^{\overline{R}} < a - \varepsilon$, except possibly for a set of probability zero. So either $X_t^{\overline{R}}$ diverges to infinity or $\liminf_{t\to\infty} X_t^{\overline{R}} = -\infty$. In the last case, ruin occurs in finite time. □

The following lemma is similar to Lemma 1.1, and so is its proof.

Lemma 2.1. *Given an initial surplus $x \geq 0$, let us consider $\overline{R} \in \Pi_x^R$, then $\delta^{\overline{R}}\left(X^{\overline{R}}_{t\wedge\tau_{\overline{R}}}\right)$ is a martingale .*

The next proposition gives the dynamic programming principle for δ and it will be used to find the HJB equation of this problem.

Proposition 2.3. *Given any initial surplus $x \geq 0$, we have that*

$$\delta(x) = \sup_{\overline{R}\in\Pi_x^R} E_x(\delta(X^{\overline{R}}_{\tau\wedge\tau^{\overline{R}}}))$$

for any stopping time τ with $P(\tau = \infty) = 0$.

Proof. It is enough to prove this proposition for the case that the stopping time is constant, that is $\tau(\omega) = T$ for all $\omega \in \Omega$. We call

$$\vartheta(x, T) = \sup_{\overline{R}\in\Pi_x^R} E_x(\delta(X^{\overline{R}}_{T\wedge\tau^{\overline{R}}})).$$

Let us prove first that $\delta(x) \leq \vartheta(x, T)$. Take any admissible strategy $\overline{R} = (R_t)_{t\geq 0} \in \Pi_x^R$. We define for any $\omega \in \Omega$ the strategy $\overline{R}_1 = (R_t^1)_{t\geq 0} \in \Pi^R_{X_T^{\overline{R}}}$ as $R_t^1(\omega) = R_{t+T}(\omega)$. The strategy $\overline{R}_1$ is admissible and

$$\begin{aligned}
\delta^{\overline{R}}(x) &= E\left(E\left(I_{\{\tau^{\overline{R}}=\infty\}}\middle|\mathcal{F}_T\right)\right)\\
&= E\left(I_{\{\tau^{\overline{R}}=\infty\}}(\delta^{\overline{R}_1}(X_T^{\overline{R}})\right)\\
&\leq E\left(I_{\{\tau^{\overline{R}}=\infty\}}(\delta(X_T^{\overline{R}})\right)\\
&\leq E_x(\delta(X^{\overline{R}}_{T\wedge\tau^{\overline{R}}})).
\end{aligned}$$

From (2.9), we get the result.

Let us prove now that $\delta(x) \geq \vartheta(x, T)$. Given any $\varepsilon > 0$, take an admissible strategy $\overline{R} = (R_t) \in \Pi_x^R$ such that

$$E_x(\delta(X^{\overline{R}}_{T\wedge\tau^{\overline{R}}})) \geq \vartheta(x, T) - \varepsilon/2. \tag{2.11}$$

Since δ is Lipschitz, we can find points $0 = x_0 < x_1 < x_2 < \cdots$ such that if $x \in [x_i, x_{i+1})$, then

$$\delta(x) - \delta(x_i) < \frac{\varepsilon}{4}. \tag{2.12}$$

Let us call $A_i = [x_i, x_{i+1})$. Take admissible strategies $\overline{R}_i = (R_t^i)_{t\geq 0} \in \Pi_{x_i}^R$ such that $\delta(x) - \delta^{\overline{R}_i}(x_i) \leq \varepsilon/4$ for all i. We define a new strategy $\overline{R}_* = (R_t^*)_{t\geq 0}$ in the following way: For $t \leq T$, take $R_t^* = R_t$, so $T \wedge \tau^{\overline{R}} = T \wedge \tau^{\overline{R}_*}$. In the case that $\tau^{\overline{R}} > T$, take i_0 such that $X^{\overline{R}}_{T\wedge\tau^{\overline{R}}} \in A_{i_0}$ and follow strategy $R^*_{T+s} = R_s^{i_0}$ for all $s \geq 0$; note that $\tau^{\overline{R}_*} \geq T + \tau^{\overline{R}_{i_0}}$. We have

$$\begin{aligned}
\delta(x) &\geq \delta^{\overline{R}*}(x) \\
&= E\left(E\left(I_{\left\{\tau^{\overline{R}*}=\infty\right\}} \middle| \mathcal{F}_T\right)\right) \\
&\geq E_x\left(\sum_i I_{A_i}(X^{\overline{R}*}_{T\wedge\tau^{\overline{R}*}})\delta^{\overline{R}_i}(x_i)\right) \\
&\geq E_x\left(\sum_i I_{A_i}(X^{\overline{R}*}_{T\wedge\tau^{\overline{R}*}})\delta(X^{\overline{R}*}_{T\wedge\tau^{\overline{R}*}})\right) - \frac{\varepsilon}{2} \\
&\geq \vartheta(x,T) - \varepsilon,
\end{aligned}$$

and so we have the result. □

We want now to derive heuristically the HJB equation. Given an initial surplus $x \geq 0$ and any fixed retained function $R \in \mathcal{R}$, we consider the constant admissible strategy $\overline{R} = (R)_{t\geq 0}$ and the controlled surplus process

$$X_t^R := x + p_R t - \sum_{i=1}^{N_t} R(U_i).$$

with ruin time τ^R. By Lemma 2.1 we have that $\delta(x) \geq E\left(\delta(X^R_{t\wedge\tau^R})\right)$ and so assuming that δ is differentiable at $x \geq 0$, we obtain from (2.7),

$$\sup_{R\in\mathcal{R}} \mathcal{L}_R(\delta)(x) \leq 0.$$

The HJB equation of the optimal problem of survival probability with reinsurance would be

$$\sup_{R\in\mathcal{R}} \mathcal{L}_R(\delta)(x) = 0. \tag{2.13}$$

In Chap. 3, we will show that δ is a viscosity solution of this equation, and in Chap. 6, we will show examples in which δ is not differentiable.

The next proposition is similar to Lemma 2.10 in [57].

Proposition 2.4. *We have that* $0 < \delta(x) < 1$ *for all* $x \geq 0$, $\lim_{x\to\infty}\delta(x) = 1$, δ *is increasing, and it is Lipschitz with Lipschitz constant* $K = \beta/\sup_{R\in\mathcal{R}} p_R$.

Proof. Take the survival probability function corresponding to the constant admissible strategy $\overline{R} = (\hat{R})_{t\geq 0}$, where $\hat{R}$ satisfies the net-profit condition. So we obtain from Lemma 1.1 that the corresponding survival probability function satisfies $\delta^{\overline{R}}(x) > 0$ for all $x \geq 0$ and that $\lim_{x\to\infty}\delta^{\overline{R}}(x) = 1$. So $\delta(x) > 0$ and $\lim_{x\to\infty}\delta(x) = 1$.

Let us prove first that $\delta < 1$; take any $\overline{R} \in \Pi_x^R$; from the proof of Proposition 2.2, we have that taking any $\varepsilon < -p_{R_0}/2$, we get that $P(X_1 \leq x - \varepsilon) \geq P_0 > 0$ where P_0 does not depend on x. So we obtain $P(\inf_{t\geq 0} X_t^{\overline{R}} < 0) \geq P_0^{(x/\varepsilon)+1}$ and so

$\delta(x) \leq 1 - P_0^{(x/\varepsilon)+1} < 1$. Let us show now that δ is nondecreasing; given $x_0 < x_1$, take any $\overline{R} = (R_t)_{t\geq 0} \in \Pi_{x_0}^R$ and consider $\overline{R}_1 = (R_t^1)_{t\geq 0} \in \Pi_{x_1}^R$ defined as $R_t^1 = R_t$ for $t \geq 0$; we obtain that $\tau^{\overline{R}} \leq \tau^{\overline{R}_1}$ and so $\delta^{\overline{R}_1}(x_1) \geq \delta^{\overline{R}}(x_0)$.

Let us prove now that δ is increasing; suppose that $\delta(x_0) = \delta(x_1)$, for any $\gamma > 0$ small enough, take $\overline{R} = (R_t)_{t\geq 0} \in \Pi_{x_0}^R$ such that $\delta^{\overline{R}}(x_0) \geq \delta(x_0) - \gamma\delta(x_0)$. Let us consider

$$\tau_{x_1} = \inf\left\{t \geq 0 : X_t^{\overline{R}} = x_1\right\}$$

and the strategy $\overline{R}_1 \in \Pi_{x_1}^R$ defined as follows: For $t \leq \tau_{x_1}$, follow strategy $\overline{R}$ and for $t > \tau_{x_1}$ follow an strategy $\overline{R}_2 \in \Pi_{2x_1-x_0}^R$ such that $\delta^{\overline{R}_2}(2x_1 - x_0) \geq \delta(2x_1 - x_0) - \gamma$. We have from Proposition 2.2 that $P(\tau_{x_1} = \infty) = 0$, so from Lemma 2.1, $\delta^{\overline{R}_1}(X_{t\wedge\tau_{x_1}}^{\overline{R}_1})$ is a martingale. Hence,

$$\delta^{\overline{R}_1}(x_1) = E_{x_1}(\delta^{\overline{R}_1}(X_{t\wedge\tau_{x_1}}^{\overline{R}_1})) = \delta^{\overline{R}_1}(2x_1 - x_0)P(\tau_{x_1} < \tau^{\overline{R}_1}).$$

Since $\delta(x_0) = \delta(x_1)$, we obtain

$$\begin{aligned}
\delta(x_0) - \gamma\delta(x_0) &\leq \delta^{\overline{R}}(x_0) \\
&= E_{x_0}(\delta^{\overline{R}}(X_{t\wedge\tau_{x_1}}^{\overline{R}})) \\
&= \delta^{\overline{R}}(x_1)P(\tau_{x_1} < \tau^{\overline{R}}) \\
&\leq \delta(x_1)P(\tau_{x_1} < \tau^{\overline{R}}) \\
&= \delta(x_0)P(\tau_{x_1} < \tau^{\overline{R}}),
\end{aligned}$$

and so $P(\tau_{x_1} < \tau^{\overline{R}}) \geq 1 - \gamma$. Then,

$$\delta(x_1) \geq \delta^{\overline{R}_1}(x_1) \geq (\delta(2x_1 - x_0) - \gamma)(1 - \gamma)$$

for any small enough $\gamma > 0$ and so $\delta(x_1) \geq \delta(2x_1 - x_0)$, but since δ is nondecreasing and $2x_1 - x_0 > x_1$ we obtain $\delta(x_0) = \delta(x_1) = \delta(2x_1 - x_0)$. Iterating this procedure and using that δ is nondecreasing we obtain $\delta(x) = \delta(x_0)$ para $x \geq x_0$, and this is a contradiction because $\delta(x_0) < 1$ and $\lim_{x\to\infty}\delta(x) = 1$.

Let us prove now that δ is Lipschitz. Take $\hat{R} \in \mathcal{R}$ such that $p_{\hat{R}} > \beta E(\hat{R}(U_i)) > 0$. Consider $x_0 \leq x_1$ and any $\varepsilon > 0$ and take an strategy $\overline{R}_1 = (R_t^1)_{t\geq 0} \in \Pi_{x_1}^R$ such that $\delta^{\overline{R}_1}(x_1) \geq \delta(x_1) - \varepsilon$. Let us define $\overline{R} = (R_t)_{t\geq 0} \in \Pi_{x_1}^R$ as follows: $R_t = \hat{R}$ for all $t \leq \tau_{x_1} = \inf\left\{t \geq 0 : X_t^{\overline{R}} = x_1\right\}$ and $R_t = R_{t-\tau_{x_1}}^1$ for $t > \tau_{x_1}$. In the event of no claims the process $X_t^{\overline{R}}$ with initial surplus x_0 reaches x_1 at time $h = (x_1 - x_0)/p_R$. So we have

$$\delta(x_0) \geq \delta^{\overline{R}}(x_0) \geq \delta^{\overline{R}}(x_1)P(h < \tau_1) = \delta^{\overline{R}}(x_1)e^{-\beta h} \geq (\delta(x_1) - \varepsilon)\, e^{-\beta h}.$$

Then, $\delta(x_0) \geq \delta(x_1)e^{-\beta h}$ and since δ is bounded by 1, we conclude that

$$0 \leq \delta(x_1) - \delta(x_0) \leq \delta(x_1)(1 - e^{-\beta(x_1 - x_0)/p_{\hat{R}}}) \leq \frac{\beta}{p_{\hat{R}}}(x_1 - x_0). \qquad \square$$

Remark 2.1. Unlike the uncontrol case, the survival probability with initial surplus zero depends on the family of retained functions $\mathcal{R}$; the natural boundary condition for the optimal survival probability function is $\lim_{x\to\infty} \delta(x) = 1$.

2.1.2 *Dividends and Reinsurance*

In this section, we consider the problem of maximizing the cumulative expected discounted dividend payouts in the case that the insurer can control the risk by reinsurance within a family of retained functions $\mathcal{R}$. We do not assume here the existence of a retained function $R \in \mathcal{R}$ with the net-profit condition $p_R > \beta E(R(U))$.

A *dividend and reinsurance strategy* is a pair $(\overline{L}, \overline{R})$ where $\overline{L} = (L_t)_{t\geq 0}$ is a dividend strategy and $\overline{R} = (R_t)_{t\geq 0}$ is a reinsurance control strategy. Given a dividend and reinsurance strategy $(\overline{L}, \overline{R})$, we define the *controlled surplus process* as

$$X_t^{\overline{L},\overline{R}} = X_t^{\overline{R}} - L_t \tag{2.14}$$

and the *ruin time* as

$$\tau^{\overline{L},\overline{R}} = \inf\{t \geq 0 : X_t^{\overline{L},\overline{R}} < 0\}.$$

The strategy $(\overline{L}, \overline{R})$ is *admissible* if the reinsurance control strategy $\overline{R}$ is admissible and the dividend strategy $\overline{L}$ is nondecreasing, càglàd, and predictable with respect to the filtration $(\mathcal{F}_t)_{t\geq 0}$ and verifies $L_0 = 0$ and

$$L_t \leq X_t^{\overline{R}} = x + \int_0^t p_{R_s} ds - \sum_{i=1}^{N_t} R_{\tau_i}(U_i)$$

for $0 \leq t < \tau^{\overline{L},\overline{R}}$. As in the case without reinsurance, we extend the definition of the admissible dividend process as $L_t = L_{\tau^{\overline{L},\overline{R}}}$ for $t \geq \tau^{\overline{L},\overline{R}}$. We denote by $\Pi_x^{L,R}$ the set of all admissible dividend and reinsurance strategies with initial surplus $x \geq 0$. Given an initial surplus $x \geq 0$ and an admissible strategy $(\overline{L}, \overline{R}) \in \Pi_x^{L,R}$, the cumulative expected discounted dividends is defined as

$$V_{\overline{L},\overline{R}}(x) = E_x\left(\int_0^{\tau^{\overline{L},\overline{R}}} e^{-cs} dL_s\right), \tag{2.15}$$

where $c > 0$ is a discount factor. The optimal value function of this problem is defined as

$$V(x) = \sup\{V_{\overline{L},\overline{R}}(x) \text{ with } (\overline{L},\overline{R}) \in \Pi_x^{L,R}\} \text{ for } x \geq 0. \tag{2.16}$$

The proofs of the following two propositions are similar to the ones of Propositions 1.2 and 1.3.

Proposition 2.5. *The optimal value function V is well defined and satisfies*

$$x + \frac{\overline{p}}{c+\beta} \leq V(x) \leq x + \frac{\overline{p}}{c} \text{ for } x \geq 0.$$

where $\overline{p} := \sup_{R \in \mathcal{R}} p_R$.

Note that the above proposition implies in particular that $V(0) \geq \overline{p}/(c+\beta) > 0$.

Proposition 2.6. *The optimal value function V is increasing and locally Lipschitz in $[0,+\infty)$ and satisfies*

$$y - x \leq V(y) - V(x) \leq \beta \frac{V(x)}{\overline{p}} (y-x)$$

for $y > x \geq 0$. So V is Lipschitz on compact sets and absolutely continuous with $1 \leq V' \leq (\beta/\overline{p})\, V$ a.e..

As in the case of optimizing dividend payments with no reinsurance, in order to obtain the HJB equation, we need to use a DPP. The proof is similar to the one of Lemma 1.2; the only difference is that here we consider admissible strategies in $\Pi_x^{L,R}$ instead of Π_x^L.

Lemma 2.2. *For any $x \geq 0$ and any stopping time $\overline{\tau}$, we can write*

$$V(x) = \sup_{(\overline{L},\overline{R}) \in \Pi_x^{L,R}} E_x\left(\int_0^{\overline{\tau} \wedge \tau^{\overline{L},\overline{R}}} e^{-cs} dL_s + e^{-c\left(\tau \wedge \tau^{\overline{L}}\right)} V(X_{\tau \wedge \tau^{\overline{L},\overline{R}}}^{\overline{L},\overline{R}}) \right).$$

Assume that V is continuously differentiable at x. Given any $l \geq 0$ and any $R \in \mathcal{R}$, let us consider the admissible strategy $(\overline{L},\overline{R}) = ((lt)_{t \geq 0}, (R)_{t \geq 0})$ which pays dividends at constant rate l and takes reinsurance with constant retained function R. Let us call the corresponding controlled surplus process $X_t^{\overline{L},\overline{R}} = X_t^{\overline{R}} - lt$ and the corresponding ruin time τ. The surplus process $X_{\tau \wedge t}^{\overline{L},\overline{R}}$ stopped at the ruin time is a Markov process, so as in (1.11) and Remark 1.7, we get

$$\tilde{\mathcal{G}}\left(X_{t \wedge \tau}^{\overline{L},\overline{R}}, V\right)(x) = \begin{cases} (p_R - l)\, V'(x) - (\beta + c)\, V(x) + \beta \mathcal{I}_R(V)(x) & \text{if } l \leq p_R \\ (p_R - l)\, V'(x) - (\beta + c)\, V(x) + \beta \mathcal{I}_R(V)(x^-) & \text{if } l > p_R, \end{cases} \tag{2.17}$$

where $\tilde{\mathcal{G}}$ is the discounted infinitesimal generator defined in (1.2) and $\mathcal{I}_R(V)$ is the operator defined in (2.8). As in (1.19), but using Lemma 2.2, we obtain the inequality

$$\sup_{l\geq 0, R\in\mathcal{R}} \left\{ l + \tilde{\mathcal{G}}\left(X_{t\wedge\tau}^{\overline{L},\overline{R}}, V\right)(x) \right\} \leq 0.$$

The HJB equation of this optimization problem is

$$\sup_{l\geq 0, R\in\mathcal{R}} \left\{ l + \tilde{\mathcal{G}}\left(X_{t\wedge\tau}^{\overline{L},\overline{R}}, V\right)(x) \right\} = 0. \tag{2.18}$$

As in Sect. 1.5.2, we obtain that the HJB equation of this problem can be rewritten as

$$\max\{1 - V'(x), \sup_{R\in\mathcal{R}} \tilde{\mathcal{L}}_R(V)(x)\} = 0, \tag{2.19}$$

where

$$\tilde{\mathcal{L}}_R(V)(x) = p_R V'(x) - (c+\beta)V(x) + \beta\mathcal{I}_R(V)(x). \tag{2.20}$$

2.2 Investments in the Classical Risk Model

In this control problems, the management of an insurance company has the possibility to invest a fraction of the surplus in the financial market. For simplicity, we assume that the claim-size distribution has bounded density. The financial market is described as a classical Black–Scholes model that consists on a risk-free asset with price process B_t and a risky asset with price process P_t satisfying

$$\begin{cases} dB_t = r_0 B_t dt \\ dP_t = rP_t dt + \sigma P_t dW_t, \end{cases}$$

where $\sigma > 0$, $r > r_0 \geq 0$, and W_t is a standard Brownian motion independent to the probability space (Ω, Σ, P) defined in (1.3); without loss of generality, we consider here $r_0 = 0$.

Let us denote by $(\Omega_3, \Sigma_3, (\mathcal{F}_t^3)_{t\geq 0}, P_3)$ the filtered probability space of the Brownian motion W_t. Let us define the filtered probability space $(\overline{\Omega}, \overline{\Sigma}, (\overline{\mathcal{F}}_t)_{t\geq 0}, \overline{P})$ as the product of probability spaces

$$(\overline{\Omega}, \overline{\Sigma}, \overline{P}) = (\Omega, \Sigma, P) \times (\Omega_3, \Sigma_3,, P_3) \tag{2.21}$$

with filtration $\overline{\mathcal{F}}_t$ generated by $\mathcal{F}_t$ and $\mathcal{F}_t^3$.

We fix the set $\Gamma \subset \mathbf{R}$ of all the fractions of the surplus which could be invested in stocks. For instance, $\Gamma = [0,1]$ means that neither short-selling of stocks nor borrowing money to buy stocks is allowed, $\Gamma = \mathbf{R}_+$ means that it is allowed borrowing money to buy stocks but short-selling of stocks is not allowed, and $\Gamma = \mathbf{R}$ means that both borrowing money and short-selling of stocks are allowed.

An investment strategy is a process $\overline{\gamma} = (\gamma_t)_{t\geq 0}$ where $\gamma_t \in \Gamma \subset \mathbf{R}$ is the fraction of the surplus invested in stocks. Given an investment strategy $\overline{\gamma} = (\gamma_t)_{t\geq 0}$, the *controlled risk process* $X_t^{\overline{\gamma}}$ should be a solution of the equation

$$X_t^{\overline{\gamma}} = x + pt - \sum_{i=1}^{N_t} U_i + \int_0^t \gamma_s X_s^{\overline{\gamma}} \left(r ds + \sigma \, d W_s \right). \tag{2.22}$$

The first three terms comes from the classical risk model and the integral term corresponds to the change of the surplus due to the investment. As before, we define the *ruin time* as

$$\tau^{\overline{\gamma}} = \inf\{t \geq 0 : X_t^{\overline{\gamma}} < 0\}.$$

An investment strategy is *admissible* if the process $(\gamma_t)_{t\geq 0}$ is predictable with respect to the filtration $\left(\mathcal{F}_t\right)_{t\geq 0}$ and there exists a unique strong solution $X_t^{\overline{\gamma}}$ of (2.22). We denote by Π_x^{γ} the set of all the admissible investment strategies with initial value x.

Remark 2.2. We introduce the process $Y_t^{\overline{\gamma}}$ as the solution of

$$Y_t^{\overline{\gamma}} = x + pt + \int_0^t \gamma_s Y_s^{\overline{\gamma}} \left(r ds + \sigma \, d W_s \right). \tag{2.23}$$

Note that $Y_t^{\overline{\gamma}}$ can be viewed as the controlled risk process without claims and that the processes $X_t^{\overline{\gamma}}$ and $Y_t^{\overline{\gamma}}$ coincide up to the first claim τ_1.

As in the case of reinsurance admissible strategies, we define a stationary investment control as the one where the investment decision depends only on the current surplus: consider a fraction $g(y) \in \Gamma$ for each $y \geq 0$, and define for any initial surplus $x \geq 0$ the surplus process $(X_t)_{t\geq 0}$ obtained by investing a fraction $g(y)$ when the current surplus is y; the process X_t should satisfy

$$X_t = x + pt - \sum_{i=1}^{N_t} U_i + \int_0^t g(X_s-) X_s \left(r ds + \sigma \, d W_s \right), \tag{2.24}$$

and it should be the controlled investment process associated with the investment strategy $(g(X_t-))_{t\geq 0}$ with initial surplus $x \geq 0$.

Definition 2.3. A *stationary investment control in* Γ is a function $g : \mathbf{R}_+ \to \Gamma$ which satisfies that the function $g(x)x$ is Lipschitz.

Remark 2.3. As is pointed out in Theorem 1.19 of Øksendal and Sulem [49], the following result holds: Given any stationary investment control g and any initial surplus $x \in \mathbf{R}_+$ there exists a unique càdlàg solution X_t^g of the stochastic integral equation (2.24). Moreover, if we define $\gamma_t = g(X_{t-}^g)$, then the strategy $\overline{\gamma}^g = (\gamma_t)_{t\geq 0} \in \Pi_x^\gamma$ and its associated surplus process $\left(X_t^{\overline{\gamma}}\right)_{t\geq 0}$ defined in 2.22 coincides with $(X_t^g)_{t\geq 0}$. The global Lipschitz condition on $g(x)x$ is only used to ensure the existence of the process $\left(X_t^g\right)_{t\geq 0}$.

Given $\gamma_0 \in \Gamma$, let us consider the constant investment strategy $\overline{\gamma} = (\gamma_0)_{t\geq 0}$; let $X_t^{\gamma_0}$ and $Y_t^{\gamma_0}$ be the processes defined in (2.22) and (2.23) and τ^{γ_0} the ruin time of the surplus process $X_t^{\gamma_0}$. Then $(X_{t\wedge\tau^{\gamma_0}}^{\gamma_0})_{t\geq 0}$, the controlled surplus process stopped at the ruin time, is a Markov process. We now compute formally its infinitesimal generator. Assume that f is twice continuously differentiable, bounded, and with bounded derivatives up to order two in $\mathbf{R}_+$ extended as $f = 0$ for $x < 0$. Defining $A_0 = \{\tau_1 > t\}$, $A_1 = \{\tau_1 \leq t, \tau_2 > t\}$ and $A_2 = \{\tau_2 \leq t\}$. We have as in Sect. 1.4, that $P(A_0) = e^{-\beta t}$, $P(A_1) = \beta t e^{-\beta t}$ and that $P(A_2) = 1 - (1+\beta t)\, e^{-\beta t} = o(t)$. So,

$$
\begin{aligned}
E\left(f(X_{t\wedge\tau^{\gamma_0}}^{\gamma_0})\right) &= E_x\left(f(X_{t\wedge\tau^{\gamma_0}}^{\gamma_0})I_{A_0}\right) + E_x\left(f(X_{t\wedge\tau^{\gamma_0}}^{\gamma_0})I_{A_1}\right) + E_x\left(f(X_{t\wedge\tau^{\gamma_0}}^{\gamma_0})I_{A_2}\right)\\
&= e^{-\beta t}\, E_x\left(f(Y_t^{\gamma_0})\right) + \beta \int_0^t \left(\int_0^\infty E_x(f(Y_s^{\gamma_0} - \alpha))dF(\alpha)\right) e^{-\beta s} ds\\
&\quad + o(t)
\end{aligned}
$$

because f is bounded. Hence,

$$
\begin{aligned}
\frac{E_x\left(f(X_{t\wedge\tau^{\gamma_0}}^{\gamma_0})\right) - f(x)}{t} &= e^{-\beta t}\left(\frac{E_x(f(Y_t^{\gamma_0})) - f(x)}{t}\right) + \frac{(e^{-\beta t}-1)}{t} f(x)\\
&\quad + \frac{\beta}{t}\int_0^t \left(\int_0^\infty E_x(f(Y_s^{\gamma_0} - \alpha))dF(\alpha)\right) e^{-\beta s} ds + \frac{o(t)}{t},
\end{aligned}
$$

and then

$$
\mathcal{G}(X_{t\wedge\tau^{\gamma_0}}^{\gamma_0}, f)(x) = \mathcal{G}(Y_t^{\gamma_0}, f)(x) - \beta f(x) + \beta \mathcal{I}(f)(x).
$$

Since f is twice continuously, we get from Itô's formula

$$
\begin{aligned}
f(Y_t) - f(x) &= \int_0^t f'(Y_s^{\gamma_0}) dY_s^{\gamma_0} + \frac{\sigma^2\gamma_0^2}{2}\int_0^t f''(Y_s^{\gamma_0})\left(Y_s^{\gamma_0}\right)^2 ds\\
&= \int_0^t \left(f'(Y_s^{\gamma_0})\left(p + r\gamma_0 Y_s^{\gamma_0}\right) + \frac{\sigma^2\gamma_0^2}{2} f''(Y_s^{\gamma_0})\left(Y_s^{\gamma_0}\right)^2\right) ds \qquad (2.25)\\
&\quad + \int_0^t f'(Y_s^{\gamma_0})\sigma\gamma_0 Y_s^{\gamma_0} dW_s.
\end{aligned}
$$

So, since the last term of (2.25) is a martingale with zero expectation, we obtain that

$$\mathcal{G}(Y_t^{\gamma_0}, f)(x) = \frac{\sigma^2\gamma_0^2 x^2}{2} f''(x) + (p + r\gamma_0 x) f'(x). \tag{2.26}$$

Therefore,

$$\mathcal{G}(X_{\tau\wedge t}^{\gamma_0}, f)(x) = \frac{\sigma^2\gamma_0^2 x^2}{2} f''(x) + (p + r\gamma_0 x) f'(x) - \beta f(x) + \beta \mathcal{I}(f)(x). \tag{2.27}$$

2.2.1 Survival Probability and Investments

Given an admissible investment strategy $\overline{\gamma}$, we define the survival probability function as

$$\delta^{\overline{\gamma}}(x) = P(\tau^{\overline{\gamma}} = \infty | X_0^{\overline{\gamma}} = x)$$

and the optimal survival probability function as

$$\delta(x) = \sup\nolimits_{\overline{\gamma}\in\Pi_x^{\gamma}} \delta^{\overline{\gamma}}(x). \tag{2.28}$$

As in Sect. 2.1.1, we have the following three results.

Proposition 2.7. *Take any admissible strategy $\overline{\gamma} \in \Pi_x^{\gamma}$, with probability one, either $X_t^{\overline{\gamma}}$ diverges to infinity as t goes to infinity or ruin occurs in finite time.*

Lemma 2.3. *Given an initial surplus $x \geq 0$, let us consider $\overline{\gamma} \in \Pi_x^{\gamma}$, then $\delta^{\overline{\gamma}}\left(X_{t\wedge\tau^{\overline{\gamma}}}^{\overline{\gamma}}\right)$ is a martingale .*

Proposition 2.8. *Given any $x \geq 0$, we have that*

$$\delta(x) = \sup_{\overline{\gamma}\in\Pi_x^{\gamma}} E_x(\delta(X_{\tau\wedge\tau^{\overline{\gamma}}}^{\overline{\gamma}}))$$

for any stopping time τ with $P(\tau = \infty) = 0$.

The argument of the proof of Proposition 2.7 is similar to the one of Proposition 2.2; the complete proof can be found in Lemma 2.18 in [57]. The proofs of Lemma 2.3 and of Proposition 2.8 are like the ones of Lemma 2.1 and Proposition 2.3.

We can now derive heuristically the HJB equation associated to this problem. Given $\gamma \in \Gamma$, consider the constant investment strategy $\overline{\gamma} = (\gamma)_{t\geq 0}$. By Proposition 2.8, we have that

$$\delta(X_{t\wedge\tau^{\overline{\gamma}}}^{\overline{\gamma}}) - \delta(x) \leq 0$$

and so $\mathcal{G}((X^{\overline{\gamma}}_{t\wedge\tau})_{t\geq 0}, \delta)(x) \leq 0$. Then by (2.27) we get

$$\mathcal{L}_\gamma(f)(x) = \tfrac{\sigma^2\gamma^2x^2}{2} f''(x) + (p + r\gamma x)\, f'(x) - \beta f(x) + \beta \mathcal{I}(f)(x) \leq 0. \quad (2.29)$$

The HJB equation of the problem of survival probability with investment is

$$\sup_{\gamma\in\Gamma} \mathcal{L}_\gamma(\delta)(x) = 0. \quad (2.30)$$

Remark 2.4. We can rewrite

$$\mathcal{L}_\gamma(f)(x) = (\tfrac{\sigma^2\gamma^2x^2}{2} f''(x) + r\gamma x f'(x)) + \mathcal{L}_0(f)(x)$$

where $\mathcal{L}_0$ is defined in (1.13).

The following proposition gives some elementary properties of the optimal survival probability function with investment; the proof is similar to the one of Proposition 2.4

Proposition 2.9. *Assume that* $0 \in \Gamma$. *We have that* $0 < \delta(x) < 1$ *for all* $x \geq 0$, $\lim_{x\to\infty} \delta(x) = 1$ *and that* δ *is Lipschitz and increasing.*

2.2.2 *Dividends and Investments*

In this section, we consider the problem of maximizing the cumulative expected discounted dividend payouts in the case that the insurer can control the risk by investing a fraction of the surplus in the financial market.

Let us fix the set $\Gamma \subset \mathbf{R}$; a *dividend and investment strategy* is a process $(\overline{L}, \overline{\gamma}) = (L_t, \gamma_t)_{t\geq 0}$ where $\overline{\gamma} = (\gamma_t)_{t\geq 0}$ is an investment strategy with $\gamma_t \in \Gamma$ and $\overline{L}$ is a dividend strategy. Given a dividend and investment strategy $(\overline{L}, \overline{\gamma})$, the *controlled risk process* $X_t^{\overline{L},\overline{\gamma}}$ is given by

$$X_t^{\overline{L},\overline{\gamma}} = x + pt + r\int_0^t X_s^{\overline{L},\overline{\gamma}} \gamma_s ds + \sigma \int_0^t \gamma_s X_s^{\overline{L},\overline{\gamma}} dW_s - \sum_{i=1}^{N_t} U_i - L_t \quad (2.31)$$

and the *ruin time* is defined as $\tau^{\overline{L},\overline{\gamma}} = \inf\{t \geq 0 : X_t^{\overline{L},\overline{\gamma}} < 0\}$. The dividend and investment strategy $(\overline{L}, \overline{\gamma})$ is *admissible* if the investment strategy $\overline{\gamma}$ is admissible and the dividend strategy $\overline{L}$ is nondecreasing, càglàd, and predictable with respect to the filtration $(\mathcal{F}_t)_{t\geq 0}$ and verifies $L_0 = 0$ and

$$L_t \leq X_t^{\overline{\gamma}} = x + pt - \sum_{i=1}^{N_t} U_i + \int_0^t \gamma_s X_s^{\overline{\gamma}} (rds + \sigma\, dW_s)$$

for $0 \leq t < \tau^{\overline{L},\overline{\gamma}}$. As in the case without investments, we extend the definition of the admissible dividend process as $L_t = L_{\tau^{\overline{L},\overline{\gamma}}}$ for $t \geq \tau^{\overline{L},\overline{\gamma}}$. We denote by $\Pi_x^{L,\gamma}$ the set of all the dividend and investment admissible strategies with initial surplus x and the value function $V_{\overline{L},\overline{\gamma}}(x)$ as the cumulative expected discounted dividends with initial surplus $x \geq 0$ that corresponds to the predictable admissible control strategy $\left(\overline{L}, \overline{\gamma}\right)$. We can write $V_{\overline{L},\overline{\gamma}}(x)$ as

$$V_{\overline{L},\overline{\gamma}}(x) = E_x(\int_0^{\tau^{\overline{L},\overline{\gamma}}} e^{-cs} dL_s). \tag{2.32}$$

where $c > 0$ is a discount factor. The optimal dividend function is defined as

$$V(x) = \sup\{V_{\overline{L},\overline{\gamma}}(x) \text{ with } \left(\overline{L}, \overline{\gamma}\right) \in \Pi_x^{L,\gamma}\} \text{ for } x \geq 0. \tag{2.33}$$

In this problem, we assume that $0 \in \Gamma$ and that

$$0 < \hat{\gamma} := \sup \Gamma < c/r \tag{2.34}$$

and show that under this assumption V is finite. We will show in Remark 2.6 that if $\Gamma = [0, \hat{\gamma}]$ with $\hat{\gamma} > c/r$, then $V(x) = \infty$ for all $x \geq 0$.

We first state some results of a related controlled continuous risk process without the downward jumps.

Lemma 2.4. *Given $x \geq 0$, any $m \in \mathbf{R}$ and any admissible investment strategy $\overline{\gamma} \in \Pi_x^{\gamma}$, consider (with a slightly abuse of notation) the process Y_t defined in (2.23), but putting m instead of p. We have that:*

(a) If $m \geq 0$, then $E_x\left(Y_t e^{-ct}\right) \leq e^{-(c-r\hat{\gamma})t}\left(x + m(1 - e^{-r\hat{\gamma}t})/\left(r\hat{\gamma}\right)\right)$.
(b) If $x > 0$ and $\tilde{\tau} = \inf\{t : Y_t < 0\}$, then $\lim_{h\to 0} P(\tilde{\tau} < h) = 0$.
(c) If $\gamma_t \equiv \gamma_0 \in \Gamma \setminus \{0\}$ for all $t \geq 0$, then

$$E_x\left(Y_t e^{-ct}\right) = e^{-(c-r\gamma_0)t}\left(x + m(1 - e^{-r\gamma_0 t})/\left(r\gamma_0\right)\right).$$

Proof. (a) Since the process

$$U_t = \exp\left(\int_0^t (r\gamma_u - \frac{\sigma^2}{2}\gamma_u^2)du + \int_0^t \sigma\gamma_u dW_u\right) \tag{2.35}$$

is the solution of the stochastic equation

$$dU_t = U_t(r\gamma_t dt + \sigma\gamma_t dW_t) \text{ with } U_0 = 1$$

and Y_t is the solution of

$$dY_t = (m + Y_t r\gamma_t)dt + \sigma\gamma_t Y_t dW_t \text{ with } Y_0 = x,$$

we can write

$$Y_t = xU_t + U_t \int_0^t mU_s^{-1}ds. \tag{2.36}$$

Let us define

$$U_{ts} = \exp\left(\int_s^t (r\gamma_u - \frac{\sigma^2}{2}\gamma_u^2)du + \int_s^t \sigma\gamma_u dW_u\right), \tag{2.37}$$

then

$$Y_t = xU_{t0} + \int_0^t mU_{ts}ds. \tag{2.38}$$

We have that

$$A_{ts} = e^{-\int_s^t r\gamma_u du}U_{ts} \tag{2.39}$$

is a martingale; see for instance Karatzas and Shreve [39]. We conclude from (2.37) to (2.39) that

$$\begin{aligned} E_x\left(Y_t e^{-ct}\right) &= E_x(e^{-ct}xe^{\int_0^t r\gamma_u du}A_{t0} + e^{-ct}\int_0^t me^{\int_s^t r\gamma_u du}A_{ts}ds) \\ &\le E_x(e^{-(c-r\hat{\gamma})t}xA_{t0} + e^{-ct}\, m\int_0^t e^{r\hat{\gamma}(t-s)}A_{ts}ds) \\ &= e^{-(c-r\hat{\gamma})t}x + e^{-(c-r\hat{\gamma})t}\, m\int_0^t e^{-r\hat{\gamma}s}ds \\ &= e^{-(c-r\hat{\gamma})t}\left(x + \tfrac{m}{r\hat{\gamma}}(1 - e^{-r\hat{\gamma}t})\right). \end{aligned}$$

(b) This result is standard for linear diffusion processes; see Borodin and Salminen [17].
(c) Follows from the proof of (a). □

Remark 2.5. Given any $(\overline{L}, \overline{\gamma}) \in \Pi_x^{L,\gamma}$, consider the controlled process $X_t^{\overline{L},\overline{\gamma}}$ and the process Y_t introduced in Lemma 2.4, with $m = p$ and investment strategy $\overline{\gamma} = (\gamma_s)_{s\ge0}$, then we have that $X_t^{\overline{L},\overline{\gamma}} \le Y_t$ for all $t \ge 0$. We can use the following argument to see this result: $X_t^{\overline{L},\overline{\gamma}} = Y_t$ for $t < \tau_1$, where τ_1 is the arrival time of the first claim, $X_{\tau_1}^{\overline{L},\overline{\gamma}} < X_{\tau_1-}^{\overline{L},\overline{\gamma}} = Y_{\tau_1}$. If there exists $t_0 \in (\tau_1, \tau_2)$ such that $X_{t_0}^{\overline{L},\overline{\gamma}} = Y_{t_0}$, then by definition $X_t^{\overline{L},\overline{\gamma}} = Y_t$ for $t \in (t_0, \tau_2)$; if this were not the case, $X_t^{\overline{L},\overline{\gamma}} < Y_t$ for $t \in (\tau_1, \tau_2)$ because the trajectories are continuous in this interval. Then $X_{\tau_2}^{\overline{L},\overline{\gamma}} < X_{\tau_2^-}^{\overline{L},\overline{\gamma}} \le Y_{\tau_2}$ and the same argument applies again.

Remark 2.6. In the case that $\Gamma = [0, \hat{\gamma}]$ with $\hat{\gamma} > c/r$, the value function V is infinite. We can assume that $x > x_0 := \left((\beta\mu - p)^+ + 1\right)/r > 0$ because, if the initial surplus x is smaller than x_0, there is a positive probability that the surplus surpasses the level x_0 (take for instance the strategy which pays no dividends and keeps all the surplus in bonds up to time $T = (x_0 - x)/p + 1$).

Given $t_0 > 0$, consider the following admissible strategy $\left(\overline{L}^{t_0}, \overline{\gamma}^{t_0}\right) \in \Pi_x^{L,\gamma}$: divide the company in two departments; one of them deals only with the investment and the payment of dividends and the other with the insurance business. The investment department starts with capital x, invests a fraction $\hat{\gamma}$ of its surplus on risky assets, and diverts to the insurance department a constant flow $p_0 = (\beta\mu - p)^+ + 1$ up to time $t_0 \wedge \tilde{\tau}_1$ when the whole surplus is paid as dividends. Here $\tilde{\tau}_1$ is the first time the surplus of the investment department reaches zero. Let $X_t^{(1)}$ be the surplus process of the investment department and Y_t be the process described in Lemma 2.4(c) with $m = -p_0$. We have that $X_{t\wedge\tilde{\tau}_1}^{(1)} = Y_t$ for $t \leq \tilde{\tau}_1$ and $X_{t\wedge\tilde{\tau}_1}^{(1)} = 0 > Y_t$ for $t > \tilde{\tau}_1$. The insurance department starts with no surplus, pays no dividends, and receives a constant flow $p_0 + p > \beta\mu$ up to time $t_0 \wedge \tilde{\tau}_1 \wedge \tilde{\tau}_2$, where $\tilde{\tau}_2$ is the ruin time of the insurance department (assuming that the insurance department keeps always receiving the constant flow $p_0 + p$). Note that $\gamma_t^{t_0} \in \Gamma$ because

$$0 \leq \gamma_t^{t_0} = \frac{\hat{\gamma} X_t^{(1)}}{X_{t-}^{\overline{L}^{t_0},\overline{\gamma}^{t_0}}} \leq \frac{\hat{\gamma} X_t^{(1)}}{X_t^{(1)}} = \hat{\gamma}$$

for $t < t_0 \wedge \tilde{\tau}_1 \wedge \tilde{\tau}_2$ and that the stopping time $\tilde{\tau}_2$ is independent of both $\tilde{\tau}_1$ and the process Y_t. Call $\tau = t_0 \wedge \tilde{\tau}_1 \wedge \tilde{\tau}_2$, the value function of this admissible strategy satisfies

$$\begin{aligned} V_{\overline{L}^{t_0},\overline{\gamma}^{t_0}}(x) &\geq E_x(X_\tau^{(1)} e^{-c\tau} I_{\{\tilde{\tau}_1 \geq t_0, \tilde{\tau}_2 \geq t_0\}}) \geq E_x(Y_{t_0} e^{-ct_0} I_{\{\tilde{\tau}_1 \geq t_0, \tilde{\tau}_2 \geq t_0\}}) \\ &= E_x(Y_{t_0} e^{-ct_0} I_{\{\tilde{\tau}_1 \geq t_0\}}) P(\{\tilde{\tau}_2 \geq t_0\}) \\ &\geq E_x(Y_{t_0} e^{-ct_0}) P(\{\tilde{\tau}_2 = \infty\}). \end{aligned}$$

As we have seen in Remark 1.3, the survival probability of the insurance department $P(\{\tilde{\tau}_2 = \infty\}) = 1 - \beta\mu/(p_0 + p) > 0$. So, from Lemma 2.4(c), we conclude that $V(x) \geq \lim_{t_0 \to \infty} V_{\overline{L}^{t_0},\overline{\gamma}^{t_0}}(x) = \infty$.

In the next two propositions, we prove that V has linear growth and we give bounds on the increments of V using the value functions of some simple admissible strategies.

Proposition 2.10. *The optimal value function V is well defined and satisfies*

$$x + p/(\beta + c) \leq V(x) \leq rx\hat{\gamma}/(c - r\hat{\gamma}) + p/(c - r\hat{\gamma}) \textit{ for } x \geq 0.$$

Proof. Consider an initial surplus $x \geq 0$. Given any $(\overline{L}, \overline{\gamma}) \in \Pi_x^{L,\gamma}$, consider the controlled process $X_t^{\overline{L},\overline{\gamma}}$ for $t \geq 0$ and define $X_t^{\overline{L},\overline{\gamma}} = 0$ for $t < 0$. Then,

$$\begin{aligned}\tilde{L}_s = L_s - \sigma \int_0^s X_u^{\overline{L},\overline{\gamma}} \gamma_u dW_u &\leq x + ps + r \int_0^s X_u^{\overline{L},\overline{\gamma}} \gamma_u du - \sum_{i=1}^{N_s} U_i \\ &\leq x + ps + r \int_0^s X_u^{\overline{L},\overline{\gamma}} \gamma_u du.\end{aligned}$$

Consider the process Y_t defined as in Lemma 2.4, with $m = p$ and the investment strategy $\overline{\gamma} = (\gamma_s)_{s \geq 0}$. Since, by Remark 2.5, we have that $X_t^{\overline{L},\overline{\gamma}} \leq Y_t$, we obtain from Lemma 2.4(a) that

$$E_x\left(X_t^{\overline{L},\overline{\gamma}} e^{-ct}\right) \leq e^{-(c-r\hat{\gamma})t}\left(x + p(1-e^{-r\hat{\gamma}t})/(r\hat{\gamma})\right).$$

Since $r\hat{\gamma} < c$ and e^{-cs} is a positive and decreasing function, we have that

$$\begin{aligned}V_{\overline{L},\overline{\gamma}}(x) &= E_x(\textstyle\int_0^\tau e^{-cs} d\tilde{L}_s) \\ &\leq E_x(\textstyle\int_0^\infty e^{-cs} d(x + ps + r \int_0^s X_u^{\overline{L},\overline{\gamma}} \gamma_u du)) \\ &\leq \textstyle\int_0^\infty e^{-cs} pds + r\hat{\gamma} \int_0^\infty E_x(e^{-cs} X_s^{\overline{L},\overline{\gamma}}) ds \\ &\leq \tfrac{p}{c} + r\hat{\gamma} \textstyle\int_0^\infty (e^{-(c-r\hat{\gamma})s}\left(x + p\frac{1-e^{-r\hat{\gamma}s}}{r\hat{\gamma}}\right)) ds \\ &= \tfrac{rx\hat{\gamma}+p}{c-r\hat{\gamma}}.\end{aligned}$$

So $V(x)$ is finite and satisfies the second inequality.

Let us prove now the first inequality. Given an initial surplus $x \geq 0$, consider the admissible strategy $(\overline{L}, 0)$ which pays immediately the whole surplus x and then pays the incoming premium p as dividends with no investment in the risky assets until the first claim, which in this strategy means ruin. Define τ_1 as the time arrival of the first claim, we have

$$V_{\overline{L},0}(x) = x + pE_x(\int_0^{\tau_1} e^{-ct}\, dt) = x + p/(\beta + c),$$

but by definition $V(x) \geq V_{\overline{L},0}(x)$, so we get the result. □

Proposition 2.11. *If $y > x \geq 0$, the function V satisfies*

(a) $V(y) - V(x) \geq y - x$
(b) $V(y) - V(x) \leq \left(e^{(c+\beta)(y-x)/p} - 1\right) V(x)$

Proof. (a) Given $\varepsilon > 0$, consider an admissible strategy $(\overline{L}, \overline{\gamma}) \in \Pi_x^{L,\gamma}$ with $V_{\overline{L},\overline{\gamma}}(x) \geq V(x) - \varepsilon$. We define a new strategy in $(\overline{L}^1, \overline{\gamma}^1) \in \Pi_y^{L,\gamma}$ in the following way: pay immediately $y - x$ as dividends and then follow the strategy $(\overline{L}, \overline{\gamma})$; this new strategy is admissible. We have that

$$V(y) \geq V_{\overline{L}^1,\overline{\gamma}^1}(y) = V_{\overline{L},\overline{\gamma}}(x) + (y - x) \geq V(x) - \varepsilon + (y - x),$$

and we obtain the result.

(b) Given $\varepsilon > 0$, take an admissible strategy $(\overline{L}, \overline{\gamma}) \in \Pi_y$ such that $V_{\overline{L},\overline{\gamma}}(y) \geq V(y) - \varepsilon$. Let us define the strategy $(\overline{L}^1, \overline{\gamma}^1) \in \Pi_x^{L,\gamma}$ that starting at x, pay no dividends and invest all the surplus in bonds if $X_t^{\overline{L}^1,\overline{\gamma}^1} < y$ and follow strategy $(\overline{L}, \overline{\gamma})$ when the current surplus reaches y. This strategy is admissible. If there is no claim up to time $t_0 = (y - x)/p$, the surplus $X_{t_0}^{\overline{L}^1,\overline{\gamma}^1} = y$. The probability of reaching y before the first claim is $e^{-\beta t_0}$, so we obtain

$$V(x) \geq V_{\overline{L}^1,\overline{\gamma}^1}(x) \geq V_{\overline{L},\overline{\gamma}}(y) e^{-(c+\beta)t_0} \geq (V(y) - \varepsilon)\, e^{-(c+\beta)(y-x)/p},$$

and we get the result. □

As a direct consequence of the previous proposition we have that V is increasing and locally Lipschitz in $[0, +\infty)$; this implies that V is absolutely continuous, that $V'(x)$ exists a.e., and that $1 \leq V'(x) \leq V(x)(c + \beta)/p$ at the points where the derivative exists.

Remark 2.7. We will prove later that the linear growth condition given by Proposition 2.10 can be improved to $V(x) \leq x + p/c$ for $x \geq 0$.

As in Lemmas 1.2 and 2.2, there is a DPP for this optimization problem.

Lemma 2.5. *For any $x \geq 0$ and any stopping time $\overline{\tau}$, we can write*

$$V(x) = \sup_{(\overline{L},\overline{\gamma}) \in \Pi_x^{L,\gamma}} E_x \left(\int_0^{\overline{\tau} \wedge \tau^{\overline{L},\overline{\gamma}}} e^{-cs} dL_s + e^{-c\left(\overline{\tau} \wedge \tau^{\overline{L},\overline{\gamma}}\right)} V(X_{\overline{\tau} \wedge \tau^{\overline{L},\overline{\gamma}}}^{\overline{L},\overline{\gamma}}) \right).$$

Assume that V is continuously differentiable at x. Given any $l \geq 0$ and any $\gamma \in \Gamma$, let us consider the admissible strategy $(\overline{L}, \overline{\gamma})$ which pays dividends at constant rate l and invest a constant fraction γ of the surplus in the financial market. Let us call the corresponding controlled surplus process $X_t^{\overline{L},\overline{\gamma}}$ and the corresponding ruin time τ. The surplus process $X_{\tau \wedge t}^{\overline{L},\overline{\gamma}}$ stopped at the ruin time is a Markov process, so by (1.14) and (2.27), we get

$$\tilde{\mathcal{G}}\left(X_{t\wedge\tau}^{\overline{L},\overline{\gamma}}, V\right)(x) = \tfrac{\sigma^2\gamma^2x^2}{2} V''(x) + (p - l + r\gamma x)\, V'(x) - (\beta + c)V(x) + \beta\mathcal{I}(V)(x). \tag{2.40}$$

Using Lemma 2.5, we have

$$V(x) \geq E_x\left(\int_0^{\tau\wedge t} e^{-cs} l \, ds\right) + E_x\left(e^{-c(\tau\wedge t)} V(X_{\tau\wedge t}^{\overline{L},\overline{\gamma}})\right),$$

and then, we obtain the inequality

$$\sup_{l\geq 0, \gamma \in \Gamma} \left\{l + \tilde{\mathcal{G}}\left(X_{t\wedge\tau}^{\overline{L},\overline{\gamma}}, V\right)(x)\right\} \leq 0.$$

The HJB equation of this optimization problem is

$$\sup_{l\geq 0, \gamma \in \Gamma} \left\{l + \tilde{\mathcal{G}}\left(X_{t\wedge\tau}^{\overline{L},\overline{\gamma}}, V\right)(x)\right\} = 0. \quad (2.41)$$

Therefore, as in Sects. 1.5.2 and 2.1.2, we obtain that this equation can be written as

$$\max\{1 - V'(x), \sup_{\gamma \in \Gamma} \tilde{\mathcal{L}}_\gamma(V)(x)\} = 0, \quad (2.42)$$

where

$$\tilde{\mathcal{L}}_\gamma(V)(x) = \tfrac{\sigma^2 \gamma^2 x^2}{2} V''(x) + (p + r\gamma x) V'(x) - (\beta + c) V(x) + \beta \mathcal{I}(V)(x). \quad (2.43)$$

2.3 Ito's Lemma and Infinitesimal Generators

The results of this section are technical and will be used to relate the composition of a function with a controlled surplus process and the corresponding infinitesimal generator. We consider nonnegative smooth enough functions $u : \mathbf{R}_+ \to \mathbf{R}$ and we extend the definition of u in $(-\infty, 0)$ as any nonnegative constant.

Proposition 2.12. Let $Z = (Z_t)_{t\geq 0}$ be the surplus process defined either in (1.1) or in (2.3) or in (2.22) with initial value x; let τ be the corresponding ruin time, then we can write for any finite stopping time $\tau^* \leq \tau$

$$u(Z_{\tau^*}) - u(x) = \int_0^{\tau^*} \mathcal{L}(u)(Z_{s-}) ds + M_{\tau^*},$$

where $\mathcal{L}$ is the operator defined either in (1.13) or in (2.7) or in (2.29). M_t is a martingale with zero expectation in the first two cases and a local martingale with zero expectation in the third case.

Proof. Let us assume first that Z is the surplus process $(X_t)_{t\geq 0}$ defined in (1.1). Take a nonnegative continuously differentiable function u in $\mathbf{R}_+$; using the change of variables formula for finite variation processes, we can write

$$
\begin{aligned}
& u(X_{\tau^*}) - u(x) \\
&= \int_0^{\tau^*} u'(X_{s^-}) dX_s + \sum_{\substack{X_{s^-} \neq X_s \\ s \le \tau^*}} \left(u(X_s) - u(X_{s^-}) - u'(X_{s^-})(X_s - X_{s^-}) \right) \\
&= \int_0^{\tau^*} u'(X_{s^-}) p ds + \sum_{\substack{X_{s^-} \neq X_s \\ s \le \tau^*}} \left(u(X_s) - u(X_{s^-}) \right) \\
&= \int_0^{\tau^*} \mathcal{L}_0(u)(X_{s^-}) ds + M^0_{\tau^*},
\end{aligned}
$$

where $\mathcal{L}_0$ is the operator defined in (1.13) and

$$
\begin{aligned}
M^0_t = & \sum_{\substack{X_{s^-} \neq X_s \\ s \le t}} \left(u(X_s) - u(X_{s^-}) \right) \\
& -\beta \int_0^t \int_0^\infty \left(u(X_{s^-} - \alpha) - u(X_{s^-}) \right) dF(\alpha) ds
\end{aligned}
\tag{2.44}
$$

is a martingale with zero expectation because $0 \le u(X_s) \le \max_{y \in [0, x+pt]} u(y)$ for $s \le t$.

In the case that Z is the surplus process $\left(X_t^{\overline{R}} \right)_{t \ge 0}$ defined in (2.3), we also have that $0 \le u(X_s^{\overline{R}}) \le \max_{y \in [0, x+pt]} u(y)$ for $s \le t$, and so we obtain a similar formula with the following zero-expectation martingale:

$$
\begin{aligned}
M_t^{\overline{R}} = & \sum_{\substack{X_{s^-} \neq X_s \\ s \le t}} \left(u(X_s^{\overline{R}}) - u(X_{s^-}^{\overline{R}}) \right) \\
& -\beta \int_0^t \int_0^\infty \left(u(X_{s^-}^{\overline{R}} - \alpha) - u(X_{s^-}^{\overline{R}}) \right) dF_{R_s}(\alpha) ds.
\end{aligned}
\tag{2.45}
$$

Finally, in the case that Z is the surplus process $\left(X_t^{\overline{\gamma}} \right)_{t \ge 0}$ defined in (2.22), take u a nonnegative twice continuously differentiable function in $\mathbf{R}_+$; we can write, using the Itô's formula,

$$
\begin{aligned}
& u(X_{\tau^*}^{\overline{\gamma}}) - u(x) \\
&= \int_0^{\tau^*} d\left(u(X_s^{\overline{\gamma}}) \right) \\
&= \int_0^{\tau^*} u'(X_{s^-}^{\overline{\gamma}}) dX_s^{\overline{\gamma}} + \int_0^{\tau^*} \frac{\left(r\gamma_s X_{s^-}^{\overline{\gamma}} \right)^2}{2} u''(X_{s^-}^{\overline{\gamma}}) ds \\
& \quad + \sum_{\substack{X_{s^-} \neq X_s \\ s \le \tau^*}} \left(u(X_s^{\overline{\gamma}}) - u(X_{s^-}^{\overline{\gamma}}) - u'(X_{s^-}^{\overline{\gamma}})(X_s^{\overline{\gamma}} - X_{s^-}^{\overline{\gamma}}) \right) \\
&= \int_0^{\tau^*} u'(X_{s^-}^{\overline{\gamma}}) \left(p + rX_{s^-}^{\overline{\gamma}} \gamma_{s^-} \right) ds + \sigma \int_0^{\tau^*} \gamma_{s^-} X_{s^-}^{\overline{\gamma}} u'(X_{s^-}^{\overline{\gamma}}) dW_s \\
& \quad + \int_0^{\tau^*} \frac{\left(r\gamma_s X_{s^-}^{\overline{\gamma}} \right)^2}{2} u''(X_{s^-}^{\overline{\gamma}}) ds + \sum_{\substack{X_{s^-} \neq X_s \\ s \le \tau^*}} \left(u(X_s^{\overline{\gamma}}) - u(X_{s^-}^{\overline{\gamma}}) \right) \\
&= \int_0^{\tau^*} \mathcal{L}_{\gamma_{s^-}}(u)(X_{s^-}^{\overline{\gamma}}) ds + M_{\tau^*}^{\gamma},
\end{aligned}
$$

where

$$M_t^\gamma = \sum_{\substack{X_{s-}\neq X_s\\ s\leq t}} \left(u(X_s^{\overline{\gamma}}) - u(X_{s-}^{\overline{\gamma}})\right) - \beta \int_0^t \int_0^\infty \left(u(X_{s-}^{\overline{\gamma}} - \alpha) - u(X_{s-}^{\overline{\gamma}})\right) dF(\alpha) ds$$
$$+\sigma \int_0^t \gamma_{s-} X_{s-}^{\overline{\gamma}} u'(X_{s-}^{\overline{\gamma}}) d W_s$$

is a local martingale with zero expectation; to see that take for instance the sequence of stopping times $\tau_n = \min\{t : X_t \leq n\}$, then $M_{t\wedge\tau_n}^\gamma$ is a martingale with zero expectation for all $n \geq 1$. Note that this local martingale is a sum of a two local martingales, one coming from the compound Poisson process (similar to M_t^0) and the other coming from the Brownian motion. □

Remark 2.8. Taking expectation in the result of Proposition 2.12, we get the Dynkin formula for the process Z (see for instance Sect. 1.3 in [49]).

Proposition 2.13. *Let* $\overline{Z} = (Z_t)_{t\geq 0}$ *be the controlled surplus process with dividends defined either in (1.7) or in (2.14) or in (2.31) with initial value* x*; let* τ *be the corresponding ruin time, then we can write for any finite stopping time* $\tau^* \leq \tau$

$$\begin{aligned} e^{-c\tau^*} u(Z_t) - u(x) = {} & \int_0^{\tau^*} \tilde{\mathcal{L}}(u)(Z_{s-}) e^{-cs} ds - \int_0^{\tau^*} e^{-cs} dL_s \\ & + \int_0^{\tau^*} (1 - u'(Z_{s-})) e^{-cs} dL_s^c \\ & + \sum_{\substack{L_{s+}\neq L_s\\ s<\tau^*}} \left(\int_0^{L_{s+}-L_s} (1 - u'(Z_s - \alpha))\, d\alpha \right) + \tilde{M}_{\tau^*} \end{aligned}$$

where $\tilde{\mathcal{L}}$ *is the operator defined either in (1.22) or in (2.20) or in (2.43).* $\tilde{M}_t$ *is a martingale with zero expectation in the first two cases and a local martingale with zero expectation in the third case.*

Proof. Let us assume first that $\overline{Z}$ is the surplus process $\left(X_t^{\overline{L}}\right)_{t\geq 0}$ defined in (1.7). Since L_t is nondecreasing and left continuous, it can be written as

$$L_t = \int_0^t dL_s^c + \sum_{\substack{X_{s+}\neq X_s\\ s<t}} (L_{s+} - L_s) \tag{2.46}$$

where L_s^c is a continuous and nondecreasing function. Take a nonnegative continuously differentiable function u in $\mathbf{R}_+$. Since the function $e^{-ct}u(x)$ is continuously differentiable in $\mathbf{R}_+$, using the expression (2.46) and the change of variables formula for finite variation processes (see for instance [51]), we can write

$$
\begin{aligned}
u(X^{\overline{L}}_{\tau^*})e^{-c\tau^*} - u(x) &= \int_0^{\tau^*} e^{-cs} d\left(u(X^{\overline{L}}_s)\right) - c\int_0^{\tau^*} u(X^{\overline{L}}_{s-})e^{-cs}ds \\
&= \int_0^{\tau^*} u'(X^{\overline{L}}_{s-})e^{-cs}pds - c\int_0^{\tau^*} u(X^{\overline{L}}_{s-})e^{-cs}ds \\
&\quad + \sum_{\substack{X^{\overline{L}}_{s-}\neq X^{\overline{L}}_s \\ s\leq\tau^*}} \left(u(X^{\overline{L}}_s) - u(X^{\overline{L}}_{s-})\right)e^{-cs} \\
&\quad - \int_0^{\tau^*} u'(X^{\overline{L}}_{s-})e^{-cs}dL^c_s + \sum_{\substack{X^{\overline{L}}_{s+}\neq X^{\overline{L}}_s \\ s<\tau^*}} \left(u(X^{\overline{L}}_{s+}) - u(X^{\overline{L}}_s)\right)e^{-cs}.
\end{aligned}
\tag{2.47}
$$

But $X^{\overline{L}}_{s+} \neq X^{\overline{L}}_s$ only at the jumps of L_s, then $X^{\overline{L}}_{s+} - X^{\overline{L}}_s = -(L_{s+} - L_s)$ and

$$
\begin{aligned}
&-\int_0^{\tau^*} u'(X^{\overline{L}}_{s-})e^{-cs}dL^c_s + \sum_{\substack{X^{\overline{L}}_{s+}\neq X^{\overline{L}}_s \\ s<\tau^*}} \left(u(X^{\overline{L}}_{s+}) - u(X^{\overline{L}}_s)\right)e^{-cs} \\
&= -\int_0^{\tau^*} u'(X^{\overline{L}}_{s-})e^{-cs}dL^c_s - \sum_{\substack{L_{s+}\neq L_s \\ s<\tau^*}} \left(\int_0^{L_{s+}-L_s} u'(X^{\overline{L}}_s - \alpha)d\alpha\right)e^{-cs}. \qquad (2.48) \\
&= -\int_0^{\tau^*} e^{-cs}dL_s + \int_0^{\tau^*} (1 - u'(X^{\overline{L}}_{s-}))e^{-cs}dL^c_s \\
&\quad + \sum_{\substack{L_{s+}\neq L_s \\ s<\tau^*}} \left(\int_0^{L_{s+}-L_s} \left(1 - u'(X^{\overline{L}}_s - \alpha)\right)d\alpha\right)e^{-cs}.
\end{aligned}
$$

On the other hand, $X^{\overline{L}}_s \neq X^{\overline{L}}_{s-}$ only at the arrival of a claim, so

$$
\begin{aligned}
\tilde{M}^0_t = &\sum_{\substack{X^{\overline{L}}_{s-}\neq X^{\overline{L}}_s \\ s\leq t}} \left(u(X^{\overline{L}}_s) - u(X^{\overline{L}}_{s-})\right)e^{-cs} \\
&-\beta\int_0^t e^{-cs}\int_0^\infty \left(u(X^{\overline{L}}_{s-} - \alpha) - u(X^{\overline{L}}_{s-})\right)dF(\alpha)ds
\end{aligned}
$$

is a martingale with zero expectation because $0 \leq u(X^{\overline{L}}_s) \leq \max_{y\in[0,x+pt]} u(y)$ for $s \leq t$. So we have the result for this case.

In the case that the controlled surplus process $\overline{Z}$ is the one defined in (2.14), the proof is similar and the corresponding zero-expectation martingale is

$$\tilde{M}_t^{\overline{R}} = \sum_{\substack{X_{s-}^{\overline{L},\overline{R}} \neq X_s^{\overline{L},\overline{R}} \\ s \leq t}} \left(u(X_s^{\overline{L},\overline{R}}) - u(X_{s-}^{\overline{L},\overline{R}}) \right) e^{-cs}$$

$$-\beta \int_0^t e^{-cs} \int_0^\infty \left(u(X_{s-}^{\overline{L},\overline{R}} - \alpha) - u(X_{s-}^{\overline{L},\overline{R}}) \right) dF_{R_x}(\alpha) ds.$$

In the case that the controlled surplus process $\overline{Z}$ is the one defined in (2.31), the corresponding zero-expectation local martingale turns to be

$$\tilde{M}_t^{\overline{\gamma}} = \sum_{\substack{X_{s-}^{\overline{L},\overline{\gamma}} \neq X_s^{\overline{L},\overline{\gamma}} \\ s \leq t}} \left(u(X_s^{\overline{L},\overline{\gamma}}) - u(X_{s-}^{\overline{L},\overline{\gamma}}) \right) e^{-cs}$$

$$-\beta \int_0^t e^{-cs} \int_0^\infty \left(u(X_{s-}^{\overline{L},\overline{\gamma}} - \alpha) - u(X_{s-}^{\overline{L},\overline{\gamma}}) \right) dF(\alpha) ds$$

$$+\sigma \int_0^t e^{-cs} \gamma_{s-} X_{s-}^{\overline{L},\overline{\gamma}} u'(X_{s-}^{\overline{L},\overline{\gamma}}) dW_s.$$

The last result can be proved with the same argument used in the previous proposition for the case of investment and in this proposition for dividend payments. □

Remark 2.9. Propositions 2.12 and 2.13 imply that $u(Z_{t\wedge\tau})$ is a martingale for any smooth solution u of the equation $\mathcal{L}(u) = 0$ and that $u(Z_{t\wedge\tau}^{\overline{L}})$ is a supermartingale for any function u which satisfies $\max\{1-u', \tilde{\mathcal{L}}(u)\} \leq 0$ in the first two cases. In the third case, the result is the same but with local martingale and local supermartingale.

2.4 Comments and References

Let us first give some references on the problem of optimal survival probability in the classical risk model. The case of reinsurance control was addressed by Schmidli [55] and Hipp and Vogt [34]. The case of investment control was considered by Hipp and Plum [32] for $\Gamma = \mathbf{R}$ and by Azcue and Muler [10] for $\Gamma = [0, \hat{\gamma}]$. Investment control with state-dependent constraints was studied by Edalati and Hipp [26]. Hipp and Taksar [33] and Schmidli [56] considered the combination of investment and reinsurance controls. In all the cases, the claim-size distributions were continuous. In this book we allow, for the problems with reinsurance control, general claim-size distributions.

Let us now give some references on optimal dividend payments in the classical risk model. The reinsurance control case was addressed in [9] and by Mnif and Sulem [47]. Azcue and Muler [11] considered the investment control problem for bounded-density claim-size distributions .

Optimal survival probability and optimal dividend payments with investment control can also be studied in the limit diffusion setting; the HJB equation for the optimal survival probability can be written as

$$\sup_{\gamma\in\Gamma}\left(\frac{\sigma^2\gamma^2x^2}{2}\delta''(x)+r\gamma x\delta'(x)\right)+\mathcal{L}_D(\delta)(x)=0, \tag{2.49}$$

(compare with Remark 2.4) and the HJB equation for the optimal dividend payments problem as

$$\max\left\{\sup_{\gamma\in\Gamma}\left(\frac{\sigma^2\gamma^2x^2}{2}V''(x)+r\gamma xV'(x)\right)+\tilde{\mathcal{L}}_D(V)(x),1-V'(x)\right\}=0, \tag{2.50}$$

where $\mathcal{L}_D$ and $\tilde{\mathcal{L}}_D$ are defined in (1.24) and (1.27), respectively. In this diffusion setting, Browne [18] studied the problem of optimal survival probability with investment control and Højgaard and Taksar [35] considered the one of dividend payments with investment control (they also included an option to reduce risk exposure by reinsurance).

Reinsurance control in the diffusion setting was considered by Schmidli [55] for optimal survival probability and by Asmussen et al. [5] for optimal dividend payments.

Some of these problems were studied in the book of Schmidli [57].

Chapter 3
Viscosity Solutions

In Chaps. 1 and 2 we have obtained heuristically the associated equations to the value functions of different control problems. We cannot expect in general to have optimal value functions smooth enough to satisfy these equations in the classical sense. In this chapter we explain the notion of viscosity solutions for ordinary integrodifferential equations and show that the optimal value functions for the classical risk model are indeed solutions of the associated equations in this weaker sense.

3.1 Examples of Non-smooth Value Functions

One of the sources for the lack of smoothness of the value functions are the discontinuities of the claim-size distribution F. Let us consider for instance the survival probability function δ defined in (1.6) and suppose that the claim-size distribution has a discontinuity at $x_0 > 0$. If δ were continuously differentiable, then it would be a solution of the differential equation (1.13) at any $x \geq 0$, so

$$\delta'(x) = \frac{\beta}{p}\delta(x) - \frac{\beta}{p}\mathcal{I}(\delta)(x);$$

but this is a contradiction because the right side of the equality is not continuous at x_0. Therefore, δ cannot be differentiable at x_0. Moreover, δ' has a downward jump at x_0 of length

$$\delta'(x_0^+) - \delta'(x_0^-) = -\frac{\beta}{p}\delta(0)\left(F(x_0) - F(x_0^-)\right) < 0.$$

To illustrate this, we consider the case where all the claims have size 1, i.e., claim-size distribution function $F(x) = I_{\{x \geq 1\}}$. The associated equation (1.13) can be written as

$$pu'(x) - \beta u(x) = 0, \text{ for } x < 1 \tag{3.1}$$

P. Azcue and N. Muler, *Stochastic Optimization in Insurance: A Dynamic Programming Approach*, SpringerBriefs in Quantitative Finance, DOI 10.1007/978-1-4939-0995-7_3,

and

$$pu'(x) - \beta u(x) + \beta u(x-1) = 0, \text{ for } x \geq 1. \tag{3.2}$$

There exists a unique continuous function u with boundary condition $u(0) = \delta(0) = (p - \beta\mu)/p$ which is a continuously differentiable solution of (3.1) in $[0, 1)$ and a continuously differentiable solution of (3.2) in $(1, \infty)$. We can obtain recursively the closed formula of u as

$$u(x) = e^{\frac{\beta}{p}x}(1 - \frac{\beta}{p}) + e^{\frac{\beta}{p}x}(1 - \frac{\beta}{p}) I_{\{x \geq 1\}} \sum_{i=1}^{[x]} e^{-\frac{\beta}{p}i} \frac{(-1)^i}{i!} \left(\frac{\beta}{p}\right)^i (x-i)^i. \tag{3.3}$$

We will show in Chap. 4 that this weak solution u is indeed the survival probability function δ; the graph of u can be seen in Sect. 6.1.1.

Even if the claim-size distribution F is continuous, the survival probability function in the problem with reinsurance defined in (2.9) can be non-differentiable. For example, consider a fixed excess of loss reinsurance contract $R(\alpha) = \min\{\alpha, a\}$ with $F(a) < 1$. Note that the claim-size distribution F_R defined in (2.1) satisfies $F_R(a) - F_R(a^-) = 1 - F(a)$, and so F_R is discontinuous at a. Assume that $p_R > \beta\mu_R$ and consider the problem of optimal survival probability with reinsurance family $\mathcal{R} = \{R\}$. The value function δ of this problem is just the survival probability of the classical risk model with premium rate p_R and claim-size distribution F_R. The associated equation (1.13) can be written as

$$p_R\delta'(x) - \beta\delta(x) + \beta\mathcal{I}_R(\delta)(x) = 0 \text{ for } x \geq 0$$

with boundary condition $\delta(0) = (p_R - \beta\mu_R)/p_R > 0$. Hence, as before, δ is not differentiable at $x_0 = a$.

The lack of regularity of the optimal value function also arises when the associated HJB equation is a fully nonlinear differential equation even in the case that F is continuous. Take, for instance, the problem of optimizing dividends introduced in (1.10), the corresponding HJB equation (1.21) has order one and we will show examples where the value function is not continuously differentiable. In the problem of optimizing dividends allowing investments presented in (2.33), the corresponding HJB equation (2.42) has order two; we will show examples in Sect. 6.2.3 where the value function is continuously differentiable but not twice continuously differentiable. Therefore, we need a weaker notion of solution for the HJB equations. We will see that the concept of viscosity solution is weak enough to allow us to prove that the value function is indeed a solution in the viscosity sense of the corresponding HJB equation, but at the same time is strong enough to characterize the value function as the unique viscosity solution under the appropriate boundary conditions. These boundary conditions include a growing condition at infinity.

3.2 Introduction to Viscosity Solutions (First Order)

In order to motivate the notion of viscosity solutions, let us study the Eikonal equation in $J = (-1, 1)$:

$$\begin{cases} L(u'(x)) = 1 - (u'(x))^2 = 0 & \text{for } x \in J \\ u(x) = 0 & \text{for } x \in \partial J. \end{cases} \tag{3.4}$$

Note that there is not a classical solution u of this equation; if this were the case, then by the mean value theorem there would be a point $c \in J$ such that $u'(c) = 0$ and this is a contradiction. However, there are infinitely many Lipschitz functions which satisfy the equation except for a zero-measure set; they are called *almost-everywhere* solutions of (3.4). Take for instance the functions u with $u(-1) = u(1) = 0$ which graphs are polygonal with slopes 1 and -1. One of these almost-everywhere solutions is the function $U(x) = 1 - |x|$ which gives the distance of the point x to the boundary of J; any other almost-everywhere function with polygonal graph has a local minimum in J. We show in Fig. 3.1a the function U in solid line together with another almost-everywhere solution in dashed line.

The method of *vanishing viscosity* can be used to select U among all the almost-everywhere solutions of the Eikonal equation. Given any $\varepsilon > 0$, consider the corresponding perturbed semilinear equation:

$$\begin{cases} \varepsilon u_\varepsilon''(x) + 1 - \left(u_\varepsilon'(x)\right)^2 = 0 & \text{for } x \in J \\ u_\varepsilon(x) = 0 & \text{for } x \in \partial J. \end{cases} \tag{3.5}$$

Using standard methods it can be proved that there exists a unique smooth solution of (3.5). Note that u_ε is an even function and cannot have a local minimum in a point $x_\varepsilon \in J$; because if this were the case, then $u_\varepsilon'(x_\varepsilon) = 0$ and $u_\varepsilon''(x_\varepsilon) \geq 0$ and so

$$\varepsilon u_\varepsilon''(x_\varepsilon) + 1 - \left(u_\varepsilon'(x_\varepsilon)\right)^2 \geq 1.$$

Hence $u_\varepsilon > 0$ and u_ε has a unique local maximum at $0 \in J$. Moreover, this solution can be written as

$$u_\varepsilon(x) = -\varepsilon \ \ln\left(\frac{e^{\frac{x}{\varepsilon}} + e^{-\frac{x}{\varepsilon}}}{e^{\frac{1}{\varepsilon}} + e^{-\frac{1}{\varepsilon}}}\right) \text{ for } x \in [-1, 1].$$

It is easy to show that u_ε converges uniformly to U as $\varepsilon \to 0^+$.

Another way to select U among the almost-everywhere solutions u of (3.4) is to require u to be solution of (3.4) at the points where it is differentiable plus some conditions on the super- and sub-differentials of the function at the points where it is not differentiable. Consider, for instance, the function u with dashed graph shown in Fig. 3.1a. At the points where u is not differentiable, there exist either lines which

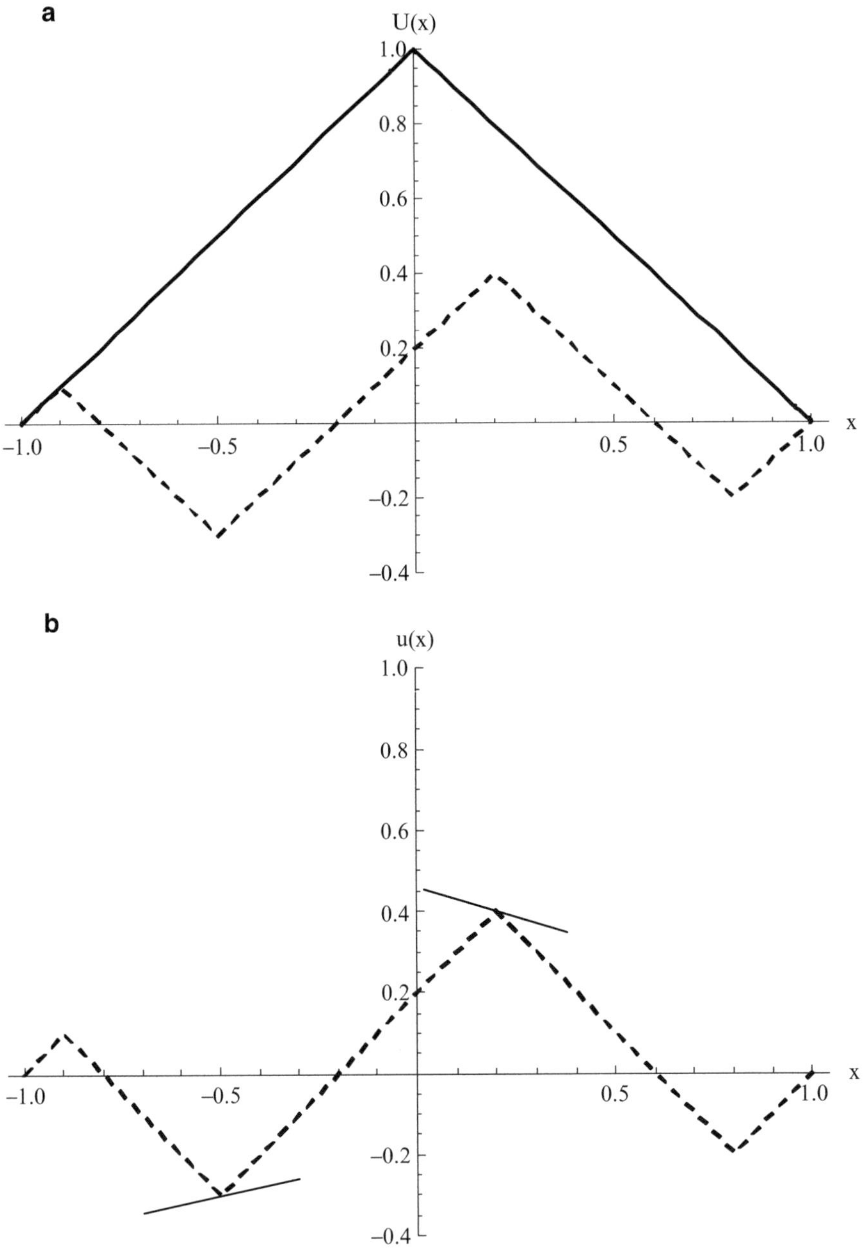

Fig. 3.1 (**a**) Almost-everywhere solutions of (3.4). (**b**) Super- and sub-differentials

are tangent from above (at the local maxima) or lines which are tangent from below (at the local minima). The slopes $\overline{d}$ of the tangent lines from above are called super-differentials of u at this point and the slopes $\underline{d}$ of the tangent lines from bellow are called sub-differential of u at this point (see Fig. 3.1b).

Let us prove that U is the only almost-everywhere solution with polygonal graph which satisfies $L(\underline{d}) \leq 0$ for all sub-differentials $\underline{d}$ and $L(\overline{d}) \geq 0$ for all super-differentials $\overline{d}$: The condition $L(\underline{d}) \leq 0$ for all sub-differentials $\underline{d}$ implies that the almost-everywhere solution u cannot have a local minimum; because if u had a local minimum, then $L(\underline{d}) = 1 - \underline{d}^2 \leq 0$ for any $\underline{d} \in [-1, 1]$ which is a contradiction. So u has to be U. Since the set of super-differentials $\overline{d}$ at the local maximum of U is $[-1, 1]$, the requirement $L(\overline{d}) = 1 - \overline{d}^2 \geq 0$ is satisfied. The functions which satisfy these conditions on the super- and sub-differentials are called viscosity solutions of the equation $L(u'(x)) = 0$. So, U is the unique viscosity solution of (3.4).

Note that if we change the sign in equation (3.4), that is

$$\begin{cases} -L(u'(x)) = (u'(x))^2 - 1 = 0 & \text{for } x \in J \\ u(x) = 0 & \text{for } x \in \partial J, \end{cases}$$

we have that the new equation has the same almost-everywhere solutions, but now the unique viscosity solution is $-U$.

We now give precise definitions of these concepts.

Definition 3.1. We say that $\overline{d}$ is a *super-differential* of u at x if

$$\limsup_{h\to 0} \frac{u(x+h) - u(x) - \overline{d}h}{|h|} \leq 0$$

and $\underline{d}$ is a *sub-differential* of u at x if

$$\liminf_{h\to 0} \frac{u(x+h) - u(x) - \underline{d}h}{|h|} \geq 0.$$

The set of all the super-differentials is denoted by $D^+(u)(x)$ and the set of all sub-differentials $D^-(u)(x)$.

Note that if $u'(x^+)$ and $u'(x^-)$ exist with $u'(x^-) < u'(x^+)$, then $D^-(u)(x) = [u'(x^-), u'(x^+)]$ and $D^+(u)(x) = \varnothing$, and if $u'(x^-) > u'(x^+)$, then $D^+(u)(x) = [u'(x^+), u'(x^-)]$ and $D^-(u)(x) = \varnothing$. Finally, if $u'(x)$ exists, then $D^+(u)(x) = D^-(u)(x) = \{u'(x)\}$.

In the case that either $u'(x^+)$ or $u'(x^-)$ do not exist, both $D^+(u)(x)$ and $D^-(u)(x)$ could be empty sets. For example, consider the locally Lipschitz function

$$u(x) = \begin{cases} 0 & \text{if } x \leq 0 \\ x^2 \sin(1/x) & \text{if } x > 0, \end{cases}$$

then

$$\lim_{h\to 0-} \frac{u(x+h)-u(x)}{|h|} = 0, \ \limsup_{h\to 0+} \frac{u(x+h)-u(x)}{|h|} = 1$$

and

$$\liminf_{h\to 0+} \frac{u(x+h)-u(x)}{|h|} = -1.$$

3.3 Viscosity Solutions of First-Order Equations

Let Z be the set of locally Lipschitz functions in $\mathbf{R}_+$. Given a function $L(x_1, x_2, x_3, g) : \mathbf{R}^3 \times Z \to \mathbf{R}$ and a domain $J \subset \mathbf{R}_+$, consider the first-order differential equations of the form

$$L(x, u(x), u'(x), u) = 0 \text{ with } x \in J. \tag{3.6}$$

The HJB equations (1.13), (1.21), (2.13), and (2.19) could be written in this form; for example, in (1.13) we have $L(x_1, x_2, x_3, g) = px_3 - \beta x_2 + \beta \mathcal{I}(g)(x_1)$ and in (2.19) we have

$$L(x_1, x_2, x_3, g) = \max\{1 - x_3, \sup_{R\in\mathcal{R}} (p_R x_3 - (c+\beta)\, x_2 + \beta \mathcal{I}_R(g)(x_1))\},$$

where $\mathcal{I}$ and $\mathcal{I}_R$ are defined in (1.12) and (2.8). In all the cases, we obtain integrodifferential equations.

Let us define the notion of viscosity solution.

Definition 3.2. A function $\overline{u} : J \to \mathbf{R}$ is a *viscosity supersolution* of the differential equation (3.6) at $x \in J$ if $\overline{u}$ is locally Lipschitz and

$$L(x, \overline{u}(x), \underline{d}, \overline{u}) \le 0$$

for all $\underline{d} \in D^-(\overline{u})(x)$. A function $\underline{u} : J \to \mathbf{R}$ is a *viscosity subsolution* of the differential equation (3.6) at $x \in J$ if $\overline{u}$ is locally Lipschitz and

$$L(x, \underline{u}(x), \overline{d}, \underline{u}) \ge 0$$

for all $\overline{d} \in D^+(\underline{u})(x)$. Finally, a function $u : J \to \mathbf{R}$ is a *viscosity solution* (3.6) at $x \in J$ if it is both viscosity subsolution and supersolution.

There is an equivalent formulation for viscosity solutions (see, for instance, Sayah [53]).

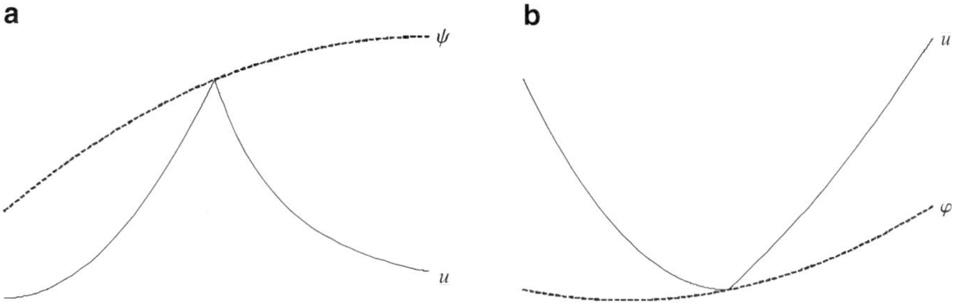

Fig. 3.2 (**a**) Test function for subsolution. (**b**) Test function for supersolution

Definition 3.3. A function $\underline{u} : J \to \mathbf{R}$ is a viscosity subsolution of (3.6) at $x \in J$ if it is locally Lipschitz and any continuously differentiable function $\psi : J \to \mathbf{R}$ with $\psi(x) = \underline{u}(x)$ and such that $\underline{u} - \psi$ reaches the maximum at x satisfies

$$L(x, \psi(x), \psi'(x), \psi) \geq 0.$$

A function $\overline{u} : J \to \mathbf{R}$ is a viscosity supersolution of (3.6) at $x \in J$ if it is locally Lipschitz and any continuously differentiable function $\varphi : J \to \mathbf{R}$ with $\varphi(x) = \overline{u}(x)$ and such that $\overline{u} - \varphi$ reaches the minimum at x satisfies

$$L(x, \varphi(x), \varphi'(x), \varphi) \leq 0.$$

If a function $u : J \to \mathbf{R}$ is both a subsolution and a supersolution at $x \in J$, it is called a viscosity solution of (3.6) at x.

As we show in Figs. 3.2a and b, the test function ψ touches $\underline{u}$ from above and the test function φ touches $\overline{u}$ from below; their derivatives $\psi'(x)$ and $\varphi'(x)$ correspond to the super- and sub-differentials at x respectively. In the first definition, the integral operator is applied to the viscosity super or subsolution, while in the second one it is applied to the test functions; the equivalence of the two definitions follows from the monotonicity and continuity of the integral operators $\mathcal{I}$ and $\mathcal{I}_R$.

Remark 3.1. The notion of classical and viscosity solution coincide at the points where the solution is differentiable. Since the viscosity solutions are locally Lipschitz, then they are differentiable and satisfy the equation (in the classical sense) almost everywhere.

We now define precisely the notion of almost-everywhere solution.

Definition 3.4. We say that $u : J \to \mathbf{R}$ is an *almost-everywhere* solution of (3.6) at $x \in J$ if u is locally Lipschitz and $L(x, u(x), u'(x), u) = 0$ at any point x where u is differentiable.

Remark 3.2. A viscosity solution of (3.6) is also an almost-everywhere solution of (3.6), as we have seen in the Eikonal example, the converse does not hold in general.

3.4 A Simple Example

We present here a toy example of equation (1.21). Let us consider this equation with $\beta = 0$, that is,

$$\max\{1 - u'(x), pu'(x) - cu(x)\} = 0 \text{ for } x > 0. \tag{3.7}$$

Any almost-everywhere solution u satisfies

$$1 \le u'(x) \le \frac{c}{p}u(x) \text{ a.e.}$$

Then $u(x) \ge p/c$ for all $x \ge 0$. So there are no almost-everywhere solutions if $u(0) < p/c$.

Gluing continuously solutions of $1-u'(x) = 0$ and solutions of $pu'(x)-cu(x) = 0$ with boundary condition $u(0) \ge p/c$, we obtain almost-everywhere solutions of (3.7). For example consider the functions

$$u_1(x) = \begin{cases} x + u(0) & \text{if } 0 \le x \le a \\ (u(0)+a)e^{\frac{c}{p}(x-a)} & \text{if } x > a \end{cases}$$

for any $a \ge 0$,

$$u_2(x) = \begin{cases} u(0)e^{\frac{c}{p}x} & \text{if } 0 \le x \le a \\ u(0)e^{\frac{c}{p}a} + x - a & \text{if } x > a \end{cases}$$

for any $a \ge 0$, and

$$u_3(x) = \begin{cases} x + u(0) & \text{if } 0 \le x \le a \\ (u(0)+a)e^{\frac{c}{p}(x-a)} & \text{if } a < x \le b \\ (u(0)+a)e^{\frac{c}{p}(b-a)} + x - b & \text{if } x > b \end{cases}$$

for any $b \ge a \ge 0$, etc. Since $1 - \left(Ae^{\frac{c}{p}x}\right)' \le 0$ and $p(x+A)' - c(x+A) \le 0$ for $A \ge p/c$, then these functions are Lipschitz solutions of (3.7) almost everywhere.

Let us find which ones of these functions are viscosity solutions with boundary condition $u(0) \ge p/c$. In the case that we glue continuously a solution of $pu'(x) - cu(x) = 0$ on the left of x_0 with a solution of $1 - u'(x) = 0$ on the right of x_0, we have that this function u is not a viscosity solution of equation (3.7) at x_0. In fact we have that $u(x_0) > u(0) \geq p/c$ and that

$$u(x) = \begin{cases} u(x_0)e^{\frac{c}{p}(x-x_0)} & \text{if } x \leq x_0 \\ u(x_0) + x - x_0 & \text{if } x > x_0 \end{cases}$$

in a neighborhood of x_0. Hence

$$u'(x_0^-) = \frac{c}{p}u(x_0) > 1, u'(x_0^+) = 1,$$

and so $D^+(u)(x_0) = [1, \frac{c}{p}u(x_0)]$. Taking $\overline{d} = \left(1 + \frac{c}{p}u(x_0)\right)/2$ we get

$$\max\left\{1 - \overline{d}, p\overline{d} - cu(x_0)\right\} < 0,$$

which implies that u is not a subsolution of (3.7) at x_0. However, if we glue continuously a solution of $1 - u'(x) = 0$ on the left of x_0 with a solution of $pu'(x) - cu(x) = 0$ on the right of x_0, the function u is indeed a viscosity solution of (3.7) at x_0. In this case

$$u(x) = \begin{cases} u(x_0) + x - x_0 & \text{if } x \leq x_0 \\ u(x_0)e^{\frac{c}{p}(x-x_0)} & \text{if } x > x_0 \end{cases}$$

in a neighborhood of x_0. Hence,

$$u'(x_0^-) = 1, u'(x_0^+) = \frac{c}{p}u(x_0) > 1,$$

and so $D^+(u)(x_0) = \varnothing$ (i.e., there is no test for subsolution) and $D^-(u)(x_0) = [1, cu(x_0)/p]$. Taking any $\underline{d} \in D^-(u)(x_0)$ we get

$$\max\left\{1 - \overline{d}, p\overline{d} - cu(x_0)\right\} \leq 0,$$

which implies that u is a supersolution of (3.7) at x_0.

We conclude that the viscosity solutions with boundary condition $u(0) \geq p/c$ are either the function $x + u(0)$ or the functions u_1 defined above. So, $x + u(0)$ is the unique viscosity solution with boundary condition $u(0)$ which satisfies a linear growth condition at infinity.

Note that equation (3.7) is the HJB equation of the problem of maximizing dividends (1.10) but without claims. It is straightforward to see that the optimal strategy with initial surplus $x \geq 0$ consists on paying immediately x as dividends and then paying the incoming premium rate as dividends forever. So, the optimal value function is

$$V(x) = x + \int_0^\infty pe^{-ct}dt = x + \frac{p}{c}.$$

This function is the unique viscosity solution with boundary condition $V(0) = p/c$ and linear growth at infinity. Moreover, the optimal value function can be characterized as the smallest viscosity solution of equation (3.7).

3.5 Value Functions Are Viscosity Solutions (First Order)

As we have explained in the previous sections, we cannot expect in general to have optimal value functions smooth enough to be classical solutions of the HJB equation. The viscosity solution is the right notion of solution for all these problems. This does not exclude that adding some assumptions on the family of claim-size distributions, the value functions become smooth enough to be classical solutions as well.

Here we show that the optimal value functions defined in (1.10) and (2.9) are viscosity solutions of the corresponding HJB equations. The optimal value function of the problem of optimizing dividends with reinsurance (defined in (2.16)) is also a viscosity solution of the corresponding HJB equation. The proof of this result is a combination of the proofs of Propositions 3.1 and 3.2; we refer to [9] for more details.

We now prove that (1.10) is a viscosity solution of (1.21). Since the claim-size distribution F is right continuous, we have the following semicontinuity result on the operator $\tilde{\mathcal{L}}_0$ defined in (1.22).

Lemma 3.1. *If u is positive and continuously differentiable, then the operator $\tilde{\mathcal{L}}_0(u)$ is right continuous and upper semicontinuous.*

Proposition 3.1. *The function V defined in (1.10) is a viscosity solution of (1.21) at any $x > 0$.*

Proof. Let us first prove that V is a viscosity supersolution of (1.21). Given an initial surplus $x_0 > 0$ and any $l \geq 0$, let us consider the admissible strategy which pays dividends at constant rate l, that is, $\overline{L} = (lt)_{t\geq 0}$. Let us call the corresponding controlled surplus process $X_t^{\overline{L}} = X_t - lt$ and the corresponding ruin time τ. Let φ be a test function for supersolution (1.21) at x_0. We extend the definition of φ as $\varphi = 0$ in $(-\infty, 0)$.

Take any $t > 0$ such that $t < x_0/(l - p)$ in the case of $p < l$ using Lemma 1.2 we get

$$\begin{aligned}\varphi(x_0) &= V(x_0)\\ &\geq E_x(\textstyle\int_0^{\tau\wedge t} e^{-c\,s}\, l\, ds) + E_x\left(e^{-c\,(\tau\wedge t)}V(X_{\tau\wedge t}^{\overline{L}})\right)\\ &\geq E_x(\textstyle\int_0^{\tau\wedge t} e^{-c\,s}\, l\, ds) + E_x\left(e^{-c\,(\tau\wedge t)}\varphi(X_{\tau\wedge t}^{\overline{L}})\right).\end{aligned}$$

So, from (1.18) we have

$$0 \geq \lim_{t \to 0+} \left(\frac{E_x(\int_0^{\tau \wedge t} e^{-cs}\, l\, ds)}{t} + \frac{E_x\left(e^{-c(\tau \wedge t)} \varphi(X^{\overline{L}}_{\tau \wedge t})\right) - \varphi(x_0)}{t} \right)$$
$$\geq l + \tilde{\mathcal{G}}\left(\left(X^l_{t \wedge \tau} \right)_{t \geq 0}, \varphi \right)(x_0).$$

Therefore, as in Sect. 1.4.2, we obtain

$$\max\{1 - \varphi'(x_0), p\varphi'(x_0) - (c + \beta)\varphi(x_0) + \beta \mathcal{I}(\varphi)(x_0)\} \leq 0,$$

where $\tilde{\mathcal{L}}_0$ is defined in (1.22).

It remains to prove that V is a viscosity subsolution at any $x > 0$ of (1.21). Arguing by contradiction, we assume that V is not a subsolution of (1.21) at $x_0 > 0$. Let us show that we can find $\varepsilon > 0$, $h \in (0, x_0/2)$ and a continuous function $\psi : \mathbf{R} \to \mathbf{R}$ such that ψ is a test function for subsolution of equation (1.21) at x_0 satisfying

$$1 - \psi'(x) \leq 0 \tag{3.8}$$

for $x \in [0, x_0 + h]$,

$$\tilde{\mathcal{L}}_0(\psi)(x) \leq -2\varepsilon c \tag{3.9}$$

for $x \in [x_0 - h, x_0 + h]$, and also

$$V(x) \leq \psi(x) - 2\varepsilon \tag{3.10}$$

for $x \in (-\infty, x_0 - h/2] \cup \{x_0 + h\}$. From Definition 3.3, if V is not a subsolution at x_0, there exist $\kappa > 0$ and a continuously differentiable function $\psi_0 : \mathbf{R}_+ \to \mathbf{R}$ such that $\psi_0(x_0) = V(x_0)$, $V - \psi_0$ reaches the maximum at x_0 in $\mathbf{R}_+$ and

$$\max\{1 - \psi_0'(x_0),\ \tilde{\mathcal{L}}_0(\psi_0)(x_0)\} < -2\kappa\beta. \tag{3.11}$$

As a first step consider $\psi_1(x) = \psi_0(x) + (\kappa/x_0^2)(x - x_0)^2$; it can be seen that ψ_1 is a continuously differentiable function satisfying $\psi_1(x_0) = V(x_0)$,

$$\psi_1(x) \geq V(x) + \frac{\kappa}{x_0^2}(x - x_0)^2 \tag{3.12}$$

for $x \geq 0$ and from (3.11) it also satisfies that

$$\max\{1 - \psi_1'(x_0),\ \tilde{\mathcal{L}}_0(\psi_1)(x_0)\} < -\kappa\beta. \tag{3.13}$$

Since ψ_1 is nonnegative and continuously differentiable, we have by Lemma 3.2 that $\tilde{\mathcal{L}}_0(\psi_1)$ is upper semicontinuous, so from (3.13) we can find $h \in (0, \min\{\sqrt{\frac{2\beta}{c+4\beta}} x_0, x_0/2)$ such that

$$\max\{1-\psi_1'(x), \tilde{\mathcal{L}}_0(\psi_1)(x)\} < -\frac{\kappa\beta}{2} \tag{3.14}$$

for $x \in [x_0 - 2h, x_0 + 2h]$. Let ϕ be an even and nonnegative twice continuously differentiable function with support included in $(-1, 1)$ such that $\int_{-1}^{1} \phi(s)ds = 1$. We define $v_n : \mathbf{R}_+ \to \mathbf{R}$ as the convolution

$$v_n(x) = \int_{-\infty}^{\infty} (\tilde{V}(x-s) + \frac{\eta h^2}{2x_0^2}) n\phi(ns)ds,$$

where $\tilde{V}$ is defined as

$$\tilde{V}(x) = \begin{cases} V(x) & \text{if } x \geq 0 \\ V(0) + x & \text{if } x \in (-V(0), 0) \\ 0 & \text{if } x \leq -V(0). \end{cases}$$

By standard techniques (see for instance Wheeden and Zygmund [67]) we have that v_n is a smooth function and v_n converges to $V + \kappa h^2 / \left(2x_0^2\right)$ uniformly on $[0, x_0 + h]$. Then, we can find n_0 large enough such that

$$V(x) + \frac{\kappa h^2}{x_0^2} \geq v_{n_0}(x) \geq V(x) + \frac{\kappa h^2}{4x_0^2} \tag{3.15}$$

for $x \in [0, x_0 + h]$. Since V is Lipschitz, we obtain by standard techniques that

$$v_{n_0}'(x) \geq 1 \tag{3.16}$$

for $x \geq 0$.

Let χ be a continuously differentiable function satisfying:

- $0 \leq \chi \leq 1$,
- $\chi(x) = 1$ for $x \in [x_0 - h, x_0 + h]$,
- $\chi(x) = 0$ for $x \notin (x_0 - 2h, x_0 + 2h)$,
- $\chi'(x) \geq 0$ for $x \in [x_0 - 2h, x_0 - h]$.

We take $\varepsilon \leq \min\left\{\kappa h^2 / \left(12x_0^2\right)), V(0)/2\right\}$ and the function ψ as

$$\psi(x) = \chi(x)\psi_1(x) + (1 - \chi(x))v_{n_0}(x) \tag{3.17}$$

for $x \geq 0$. We extend the definition of ψ as $\psi(x) = V(0)$ for $x < 0$.

It can be shown from (3.12), (3.14), (3.15), (3.16), and (3.17) that the function ψ satisfies $\psi(x_0) = V(x_0)$, (3.8), (3.9), and (3.10). Since ψ is continuously differentiable we can find a positive constant C such that

$$\tilde{\mathcal{L}}_0(\psi)(x) \leq C \tag{3.18}$$

for all $x \in [0, x_0 + h]$. Take

$$0 < \theta < \min \left\{ \frac{\varepsilon}{2\beta C}, \frac{1}{4c} \right\}. \tag{3.19}$$

Let us take any admissible strategy $\overline{L} = (L_t)_{t \geq 0} \in \Pi^L_{x_0}$, define X_t as the corresponding controlled risk process starting at x_0, and define

$$\overline{\tau} = \inf\{t > 0 : \ X_t \geq x_0 + h\}, \underline{\tau} = \inf\{t > 0 : \ X_t \leq x_0 - h\}$$

and

$$\tau^* = \overline{\tau} \wedge (\underline{\tau} + \theta) \wedge \tilde{\tau} \wedge \tau^{\overline{L}}.$$

Noting that τ^* is less than or equal to the ruin time of X_t, it is easy to see that τ^* is finite for h small enough, take for example h such that $F(h) < 1$. It is necessary to introduce $\theta > 0$ because the value $X_{\underline{\tau}}$ corresponds to the surplus before a possible lump dividend payment and so it could happen that $X_{\underline{\tau}} > x_0 - h$ and $X_{\underline{\tau}^+} \leq x_0 - h$.

From (3.10), the third inequality of (3.19), we get that

$$V(X_{\tau^*}) \leq \psi(X_{\tau^*}) - 2\varepsilon. \tag{3.20}$$

By Proposition 2.13 and (3.8), we obtain

$$\psi(X_{\tau^*})e^{-c\tau^*} - \psi(x_0) \leq \int_0^{\tau^*} \tilde{\mathcal{L}}_0(\psi)(X_{s^-})e^{-cs}ds - \int_0^{\tau^*} e^{-cs}dL_s + \tilde{M}_{\tau^*}, \tag{3.21}$$

where $\tilde{M}_t$ is a zero-expectation martingale. Using the second inequality of (3.8), (3.18) and the first inequality of (3.19), we get

$$\begin{aligned} \textstyle\int_0^{\tau^*} \tilde{\mathcal{L}}_0(\psi)(X_{s^-})e^{-cs}ds &= \textstyle\int_0^{\overline{\tau}\wedge\underline{\tau}} \tilde{\mathcal{L}}_0(\psi)(X_{s^-})e^{-cs}ds \\ &\quad + \textstyle\int_{\overline{\tau}\wedge\underline{\tau}}^{\tau^*} \tilde{\mathcal{L}}_0(\psi)(X_{s^-})e^{-cs}ds \\ &\leq -2\varepsilon c \textstyle\int_0^{\overline{\tau}\wedge\underline{\tau}} e^{-cs}ds + \frac{\varepsilon}{2}. \end{aligned} \tag{3.22}$$

We also have, from the second inequality of (3.19),

$$\begin{aligned} 2\varepsilon c \textstyle\int_0^{\overline{\tau}\wedge\underline{\tau}} e^{-cs}ds &= 2\varepsilon c \textstyle\int_0^{\tau^*} e^{-cs}ds - 2\varepsilon c \textstyle\int_{\overline{\tau}\wedge\underline{\tau}}^{\tau^*} e^{-cs}ds \\ &\geq 2\varepsilon c \textstyle\int_0^{\tau^*} e^{-cs}ds - \frac{\varepsilon}{2}. \end{aligned} \tag{3.23}$$

From (3.20), using (3.21), (3.22), and (3.23), it follows that

$$\begin{aligned} e^{-c\tau^*}V(X_{\tau^*}) &\leq \left(\psi(x_0) - e^{-c\tau^*}2\varepsilon\right) + (\psi(X_{\tau^*})e^{-c\tau^*} - \psi(x_0)) \\ &\leq \psi(x_0) - e^{-c\tau^*}2\varepsilon - 2\varepsilon c \textstyle\int_0^{\tau^*} e^{-cs}ds + \varepsilon \\ &\quad - \textstyle\int_0^{\tau^*} e^{-cs}dL_s + M_{\tau^*}. \end{aligned} \tag{3.24}$$

Since

$$E_{x_0}\left(1-e^{-c\tau^*}\right)=cE_{x_0}\left(\int_0^{\tau^*}e^{-cs}ds\right),$$

we get from (3.24), using Lemma 1.2,

$$V(x_0)=\sup_{\overline{L}\in\Pi_x^L}E_{x_0}\left(\int_0^{\tau^*}e^{-cs}dL_s+e^{-c\tau^*}V(X_{\tau^*}^{\overline{L}})\right)\le\psi(x_0)-\varepsilon$$

and this contradicts the assumption that $V(x_0)=\psi(x_0)$. □

We now prove that (2.9) is a viscosity solution of (2.13). As in the case of dividends, we first need to show a semicontinuity result on the operator $\sup_{R\in\mathcal{R}}\mathcal{L}_R$. Since the operator is actually a supremum of operators on the family of reinsurance we considered, the proof of this result is not so direct as the proof of Lemma 3.1.

Lemma 3.2. *Consider $\mathcal{R}$ as one of the reinsurance families $\mathcal{R}_A$, $\mathcal{R}_{XL}$, $\mathcal{R}_P$, and $\mathcal{R}_F$ introduced in Definition 2.1. If u is a nonnegative and a twice continuously differentiable function defined in $\mathbf{R}_+$ (extended as $u(x)=0$ for $x<0$), then $H=\sup_{R\in\mathcal{R}}\mathcal{L}_R(u)$ is upper semicontinuous and right continuous for $x>0$. Moreover, we have that for any $A>0$ and $h\in(0,1)$, there exists $K_A>0$ such that*

$$|H(y)-H(x_0)|\le K_A\left(y-x_0+F(y)-F(x_0)\right)$$

for all the reinsurance families considered and $0\le x_0,y\le A$.

Proof. Let us prove first that H is left upper semicontinuous. Given $0\le y<x_0\le A$, take reinsurance policies $R_y\in\mathcal{R}$ such that

$$\sup_{R\in\mathcal{R}}\mathcal{L}_R(u)(y)\le\mathcal{L}_{R_y}(u)(y)+(x_0-y). \tag{3.25}$$

Since u is continuously differentiable, we have

$$\limsup_{y\to x_0^-}\mathcal{L}_{R_y}(u)(x_0)\ge\limsup_{y\to x_0^-}\mathcal{L}_{R_y}(u)(y). \tag{3.26}$$

From (3.25) and (3.26), we get

$$H(x_0)=\sup_{R\in\mathcal{R}}\mathcal{L}_R(u)(x_0)\ge\limsup_{y\to x_0^-}\mathcal{L}_{R_y}(u)(x_0)\ge\limsup_{y\to x_0^-}H(y).$$

We prove now that H is right lower semicontinuous, and there exists $K_A>0$ such that

$$H(x_0)-H(y)\le K_A\left(y-x_0+F(y)-F(x_0)\right)$$

for any $0 < x_0 < y \le A$. Given any $R_{x_0} \in \mathcal{R}$ and any $y > x_0$, we have that

$$\begin{aligned} &\mathcal{L}_{R_{x_0}}(x_0) - \mathcal{L}_{R_{x_0}}(y) \\ &\le p\left|u'(x_0) - u'(y)\right| + \beta\left|u(x_0-\alpha) - u(y-\alpha)\right| \\ &\le p\,\sup_{x\in[0,A]}\left|u''(x)\right|(y-x_0) + \beta\,\sup_{x\in[0,A]}\left|u'(x)\right|(y-x_0). \end{aligned} \tag{3.27}$$

So, if we take $R_{x_0} \in \mathcal{R}$ such that $\sup_{R\in\mathcal{R}} \mathcal{L}_R(x_0) - \mathcal{L}_{R_{x_0}}(x_0) < \varepsilon$, we get

$$\begin{aligned} &\sup_{R\in\mathcal{R}} \mathcal{L}_R(x_0) - \sup_{R\in\mathcal{R}} \mathcal{L}_R(y) \\ &\le \mathcal{L}_{R_{x_0}}(x_0) - \mathcal{L}_{R_{x_0}}(y) + \varepsilon \\ &\le \left(p\,\sup_{x\in[0,A]}\left|u''(x)\right| + \beta\,\sup_{x\in[0,A]}\left|u'(x)\right|\right)(y-x_0) + \varepsilon. \end{aligned}$$

Therefore,

$$H(x_0) - H(y) \le \left(p \sup_{x\in[0,A]}\left|u''(x)\right| + \beta \sup_{x\in[0,A]}\left|u'(x)\right|\right)(y-x_0). \tag{3.28}$$

Let us prove now that H is right upper semicontinuous and there exists $K_A > 0$ such that

$$H(y) - H(x_0) \le K_A\left(y - x_0 + F(y) - F(x_0)\right) \tag{3.29}$$

for any $0 < x_0 < y \le A$. We prove this result for $\mathcal{R}_{XL}$, the proof for the case $\mathcal{R}_A$ is given in Lemma A.1 of [9] and the proofs for $\mathcal{R}_A$ and $\mathcal{R}_F$ are simpler.

Given any $\varepsilon > 0$, consider $0 \le x_0 < y \le A$ and $R_y \in \mathcal{R}_{XL}$ such that $H(y) - \mathcal{L}_{R_y}(y) \le \varepsilon$. Let us define

$$\overline{R}_y(\alpha) = \begin{cases} R_y(\alpha) & \text{if } R_y(\alpha) = \alpha \text{ for all } \alpha \\ R_y(\alpha) & \text{if } R_y(\alpha) = \alpha \wedge a_y \text{ with } a_y \notin (x_0, y) \\ \alpha \wedge \left(a_y \wedge x_0\right) & \text{if } R_y(\alpha) = \alpha \wedge a_y \text{ for all } \alpha \text{ with } a_y \in (x_0, y), \end{cases}$$

where a_y is the retention level of R_y. Note that

$$R_y(\alpha) - (y - x_0) \le \overline{R}_y(\alpha) \le R_y(\alpha)$$

and so

$$p_{Ry} - (1+\eta_1)\beta(y-x_0) \le p_{\overline{R}y} \le p_{Ry}.$$

We can take y close enough to x_0 in such a way that $p_{Ry} - (1+\eta_1)\beta(y-x_0) > 0$ and so $p_{\overline{R}y} > 0$. Then if $a_y \in [0, y)$, we get

$$\int_0^\infty \left(u(y - R_y(\alpha)) - u(x_0 - \overline{R}_y(\alpha))\right) dF(\alpha) \le \sup_{x\in[0,A]}\left|u'(x)\right|(y-x_0)$$

and if $a_y \geq y > x_0$,we obtain

$$\begin{aligned}&\int_0^\infty \left(u(y - R_y(\alpha)) - u(x_0 - \overline{R}_y(\alpha))\right) dF(\alpha)\\&\leq \sup_{x\in[0,A]} |u'(x)| \, (y - x_0) + \sup_{x\in[0,A]} |u(x)| \, (F(y) - F(x_0)).\end{aligned}$$

Hence,

$$\begin{aligned}&H(y) - H(x_0)\\&\leq \mathcal{L}_{R_y}(y) - \mathcal{L}_{\overline{R}y}(x_0) + \varepsilon\\&\leq p_{R_y} \sup_{x\in[0,A]} |u''(x)| \, (y - x_0) + (1 + \eta_1)\beta(y - x_0)u'(x_0)\\&\quad +\beta \sup_{x\in[0,A]} |u'(x)| \, (y - x_0) + \beta \sup_{x\in[0,A]} |u'(x)| \, (y - x_0)\\&\quad +\beta(F(y) - F(x_0)) \sup_{x\in[0,A]} |u(x)| + \varepsilon,\end{aligned} \tag{3.30}$$

and so we get that H is right upper semicontinuous and (3.29). □

Let us show now the main result.

Proposition 3.2. *Consider $\mathcal{R}$ as one of the reinsurance families $\mathcal{R}_A$, $\mathcal{R}_{XL}$, $\mathcal{R}_P$, and $\mathcal{R}_F$ introduced in Definition 2.1. The function δ defined in (2.9) is a viscosity solution of (2.13) in each of the reinsurance families.*

Proof. Let us prove first that δ is a viscosity supersolution. Given an initial surplus $x_0 > 0$ and any fixed retained function $R \in \mathcal{R}$, we consider the constant admissible strategy $\overline{R} = (R)_{t\geq 0}$ and the controlled surplus process

$$X_t^{\overline{R}} := x_0 + p_R t - \sum_{i=1}^{N_t} R(U_i)$$

with ruin time τ. By Lemma 2.1, we have $\delta(x_0) \geq E\left(\delta(X_{t\wedge\tau}^{\overline{R}})\right)$. Let φ be a test function for supersolution of (2.13) and let us extend as $\varphi(x) = 0$ for $x < 0$. We get

$$\varphi(x_0) = \delta(x_0) \geq E\left(\delta(X_{t\wedge\tau}^{\overline{R}})\right) \geq E\left(\varphi(X_{t\wedge\tau}^{\overline{R}})\right).$$

So, $E\left(\varphi(X_{t\wedge\tau}^{\overline{R}}) - \varphi(x_0)\right) \leq 0$, and since φ is continuously differentiable at x_0, we obtain from (2.7) that

$$\mathcal{L}_R(\varphi)(x_0) = \mathcal{G}(X_{t\wedge\tau}^{\overline{R}}, \varphi)(x_0) = \lim_{t\to 0+} \frac{E\left(\varphi(X_{t\wedge\tau}^{\overline{R}}) - \varphi(x_0)\right)}{t} \leq 0$$

and then

$$\sup_{R\in\mathcal{R}} \mathcal{L}_R(\varphi)(x_0) \leq 0.$$

It remains to prove that δ is a viscosity subsolution. Arguing by contradiction, we assume that δ is not a subsolution of (2.13) at x_0 with $x_0 > 0$. With a similar argument given in the proof of Proposition 3.1, we can find $\varepsilon > 0$, $h \in (0, x_0/2)$ and a continuous function $\psi : \mathbf{R} \to \mathbf{R}$ with $\psi(x) = \delta(0)$ for $x < 0$ such that ψ is a test function for subsolution of equation (2.13) at x_0 satisfying

$$\sup_{R \in \mathcal{R}} \mathcal{L}_R(\psi)(x) \leq 0 \tag{3.31}$$

for $x \in [x_0 - h, x_0 + h]$ and also

$$\delta(x) \leq \psi(x) - \varepsilon \tag{3.32}$$

for $x \in (-\infty, x_0 - h] \cup \{x_0 + h\}$. Let us take any admissible strategy $\overline{R} = (R_t)_{t \geq 0} \in \Pi^R_{x_0}$ with $R_t \in \mathcal{R}$, define $(X_t)_{t \geq 0}$ as the corresponding controlled risk process starting at x_0 and define the stopping time

$$\tau^* = \inf\{t > 0 : \ X_t \notin [x_0 - h, x_0 + h]\}.$$

We have that

$$\delta(X_{\tau^*}) \leq \psi(X_{\tau^*}) - \varepsilon. \tag{3.33}$$

The function $\psi(x)$ is continuously differentiable, so using Proposition 2.12, we obtain

$$\psi(X_{\tau^*}) - \psi(x_0) = \int_0^{\tau^*} \mathcal{L}_{R_s}(\psi)(X_{s^-}) ds + M_{\tau^*}, \tag{3.34}$$

where M_t is a zero-expectation martingale. From (3.31), we get that $\mathcal{L}_{R_s}(\psi)(x) \leq 0$ for $R_s \in \mathcal{R}$ and $x \in [x_0 - h, x_0 + h]$. Since $X_{s^-} \subset [x_0 - h, x_0 + h]$ for all $s \leq \tau^*$, we obtain

$$\int_0^{\tau^*} \mathcal{L}_{R_s}(\psi)(X_{s^-}) ds \leq 0. \tag{3.35}$$

From (3.34) and (3.35), we get $\psi(X_{\tau^*}) - \psi(x_0) \leq M_{\tau^*}$ and so from (3.33) and $\psi(x_0) = \delta(x_0)$ we obtain

$$E_{x_0}(\delta(X_{\tau^*})) \leq E_{x_0}(\psi(X_{\tau^*})) - 2\varepsilon \leq \psi(x_0) - \varepsilon = \delta(x_0) - \varepsilon. \tag{3.36}$$

So we get from (3.36) and using Proposition 2.3 that

$$\delta(x_0) = \sup_{\overline{R} \in \Pi^R_{x_0}} E_{x_0}(\delta(X_{\tau^*})) \leq \delta(x_0) - \varepsilon$$

which is a contradiction. □

Remark 3.3. In the particular case that the family $\mathcal{R}$ consists in just the retention function $R_I(\alpha) = \alpha$ (no reinsurance), the last proposition proves that the survival probability function δ defined in (1.6) is a viscosity solution of (1.13).

3.6 Viscosity Solutions (Second Order)

The notion of viscosity solutions for second-order differential equations is similar to the one of first order. Given now a function $L(x_1, x_2, x_3, x_4, g) : \mathbf{R}^4 \times \mathbf{Z} \to \mathbf{R}$ and a domain $J \subset \mathbf{R}_+$, consider the second-order differential equations of the form

$$L(x, u(x), u'(x), u''(x), u) = 0 \text{ with } x \in J. \tag{3.37}$$

The HJB equations (2.30) and (2.42) could be written in this form; for example, in (2.30) we have

$$L(x_1, x_2, x_3, x_4, g) = \sup_{\gamma \in \Gamma} \left\{ \frac{\sigma^2 x_1^2 \gamma^2}{2} x_4 + (p + r x_1 \gamma) x_3 - \beta x_2 + \beta \mathcal{I}(g)(x_1) \right\},$$

Let us state the definition (similar to Definition 3.3) of viscosity solution of second-order integrodifferential operators.

Definition 3.5. A function $\underline{u} : J \to \mathbf{R}$ is a *viscosity subsolution* of (3.37) at $x \in J$ if it is locally Lipschitz and any continuously differentiable function $\psi : J \to \mathbf{R}$ with $\psi(x) = \underline{u}(x)$ and such that $\underline{u} - \psi$ reaches the maximum at x satisfies

$$L(x, \psi(x), \psi'(x), \psi''(x), \psi) \geq 0;$$

a function $\overline{u} : J \to \mathbf{R}$ is a *viscosity supersolution* of (3.37) at $x \in J$ if it is locally Lipschitz and any continuously differentiable function $\varphi : J \to \mathbf{R}$ with $\psi(x) = \overline{u}(x)$ and such that $\overline{u} - \varphi$ reaches the minimum at x satisfies

$$L(x, \varphi(x), \varphi'(x), \varphi''(x), \varphi) \leq 0.$$

If a function $u : J \to \mathbf{R}$ is both a subsolution and a supersolution at $x \in J$, it is called a *viscosity solution* of (3.37) at x.

Let us state now an equivalent definition (analogous to Definition 3.2). The proof of the equivalence of these definitions is standard; see, for instance, Benth, Karlsen, and Reikvam [15]. We use both definitions indistinctly. First, we define the notion of second sub- and super-differentials that generalizes the one given in Definition 3.1 for first-order differential equations.

Definition 3.6. We say that $(\overline{d}, \overline{q})$ is a *second-order super-differential* of u at x if

$$\limsup_{h\to 0} \frac{u(x+h) - u(x) - h\overline{d} - h^2\overline{q}/2}{h^2} \leq 0$$

and $\left(\underline{d}, \underline{q}\right)$ is a *second-order sub-differential* of u at x if

$$\liminf_{h\to 0} \frac{u(x+h) - u(x) - h\underline{d} - h^2\underline{q}/2}{h^2} \geq 0.$$

The set of all the super-differentials is denoted by $D_2^+(u)(x)$ and the set of all sub-differentials $D_2^-(u)(x)$.

Note that if $u'(x^+)$, $u'(x^-)$, $u''(x^+)$, and $u''(x^-)$ exist, the set of sub- and super-differentials are easy to characterize.

If $u'(x^-) < u'(x^+)$, then $D_2^+(u)(x) = \varnothing$ and $\left(\underline{d}, \underline{q}\right) \in D_2^-(u)(x)$ satisfy either $\underline{d} \in (u'(x^-), u'(x^+))$ and $\underline{q} \in \mathbf{R}$, or $\underline{d} = u'(x^-)$ and $\underline{q} \leq u''(x^-)$, or $\underline{d} = u'(x^+)$ and $\underline{q} \leq u''(x^+)$.

If $u'(x^+) < u'(x^-)$, then $D_2^-(u)(x) = \varnothing$ and $\left(\overline{d}, \overline{q}\right) \in D_2^+(u)(x)$ satisfy either $\overline{d} \in (u'(x^+), u'(x^-))$ and $\overline{q} \in \mathbf{R}$, or $\overline{d} = u'(x^-)$ and $\overline{q} \geq u''(x^-)$, or $\overline{d} = u'(x^+)$ and $\overline{q} \geq u''(x^-)$.

If $u'(x)$ exists, then

$$D_2^-(u)(x) = \left\{(u'(x), \underline{q}) : \underline{q} \leq u''(x^-) \wedge u''(x^+)\right\}$$

and

$$D_2^+(u)(x) = \left\{(u'(x), \overline{q}) : \overline{q} \geq u''(x^-) \vee u''(x^+)\right\}.$$

Definition 3.7. A function $\overline{u} : J \to \mathbf{R}$ is a *viscosity supersolution* of the differential equation (3.37) at $x \in J$ if $\overline{u}$ is locally Lipschitz and

$$L(x, \overline{u}(x), \underline{d}, \underline{q}, \overline{u}) \leq 0$$

for all $\left(\underline{d}, \underline{q}\right) \in D_2^-(\overline{u})(x)$; a function $\underline{u} : J \to \mathbf{R}$ is a *viscosity subsolution* of the differential equation (3.37) at $x \in J$ if $\overline{u}$ is locally Lipschitz and

$$L(x, \underline{u}(x), \overline{d}, \overline{q}, \underline{u}) \geq 0$$

for all $\left(\overline{d}, \overline{q}\right) \in D_2^+(\underline{u})(x)$. Finally, a function $u : J \to \mathbf{R}$ is a *viscosity solution* (3.37) at $x \in J$ if it is both viscosity subsolution and supersolution.

The pairs $\left(\underline{d},\underline{q}\right) \in D_2^-(\overline{u})(x)$ correspond to the derivatives $(\varphi'(x), \varphi''(x))$ of the test functions for supersolutions in Definition 3.5 and the pairs $\left(\overline{d},\overline{q}\right) \in D_2^+(\overline{u})(x)$ correspond to the derivatives $(\psi'(x), \psi''(x))$ of the test functions for supersolutions in Definition 3.5. Note that if a function u is a viscosity solution of (3.37) and has first and second derivatives at a point x, it is a classical solution of (3.37) at x.

The definition of viscosity solution works for operators $L(x_1, x_2, x_3, x_4, g)$ that are nondecreasing on the variable x_4 since if $\left(\underline{d_0}, \underline{q_0}\right) \in D_2^-(\overline{u})(x)$, then $\left(\underline{d_0}, \underline{q}\right) \in D_2^-(\overline{u})(x)$ for all $\underline{q} \leq \underline{q_0}$ (and also if $\left(\overline{d}_0, \overline{q}_0\right) \in D_2^+(\overline{u})(x)$, then $\left(\overline{d}_0, \overline{q}\right) \in D_2^+(\overline{u})(x)$ for all $\overline{q} \geq \overline{q}_0$).

3.7 Semiconcavity

Consider the simple ordinary differential equation $u'' = 0$. It can be easily proved that the viscosity subsolutions are the convex functions and the viscosity supersolutions are the concave functions. For instance, the function $|x|$ is a viscosity supersolution of this equation because the second derivative is zero at all points except at $x = 0$ and there is no test function for viscosity subsolution (touching form above) at $x = 0$. So, the viscosity solutions are the affine functions which are also classical solutions. There are many more almost-everywhere locally Lipschitz solutions of this equation, for instance, all the continuous functions with polygonal graphs.

Let us now define the notion of semiconcavity. This is an important property of the viscosity solutions of second-order differential equations.

Definition 3.8. A function u is *semiconcave* in an interval $[x_0, x_1]$ if there exists $K \geq 0$ such that $u(x) - Kx^2/2$ is concave in this interval.

As above, it can be seen that the semiconcave functions with concave constant K are the viscosity supersolutions of the equation

$$u'' = K.$$

Remark 3.4. Any semiconcave function u has first and second derivatives a.e. with $u'' \leq K$ a.e. However, there exist semiconcave functions u such that u' is continuous but not absolutely continuous, take for example

$$u(x) = -\int_0^x C(s)ds$$

where C is the Cantor function.

3.8 Value Functions Are Viscosity Solutions (Second Order)

We show in this section that the optimal dividend payments and the optimal survival probability functions defined in (2.33) and in (2.28) are viscosity solutions of the corresponding second-order HJB equations. We assume that Γ is bounded and the claim-size distribution F is continuous.

The next proposition states that the optimal value function of the dividend payments problem defined in (2.33) is a viscosity solution of (2.42). The proof is similar to the one of Proposition 3.1. In order to prove that the value function is a supersolution one should use the infinitesimal generator (2.40) instead of (2.17); for the proof of subsolution, one has to use the result of Proposition 2.13 for the process (2.31).

Proposition 3.3. *The function V defined in (2.33) is a viscosity solution of (2.42) at any $x > 0$.*

Let us show now that any viscosity supersolution of the HJB equation (2.42) is semiconcave and so V is semiconcave.

Proposition 3.4. *Any nondecreasing supersolution $\overline{u}$ of equation (2.42) in $(0,\infty)$ is semiconcave in any interval $[x_0, x_1] \subset (0,\infty)$.*

Proof. In this proof we only use that $\overline{u}$ is a viscosity supersolution of $\tilde{\mathcal{L}}_\gamma = 0$ for some $\gamma \in \Gamma - \{0\}$. It is enough to prove that there exists a constant K and a sequence of semiconcave functions v_n in $[0, x_1]$ such that $v_n'' \leq K$ a.e. and $v_n \to \overline{u}$ uniformly in $[0, x_1]$.

Since $\overline{u}$ is an absolutely continuous function, there exists $k_0 \geq 1$ such that

$$|\overline{u}(x) - \overline{u}(y)| \leq k_0 \, |x - y|$$

for all $x, y \in [0, x_1]$. Let us define for any $x \in [0, x_1]$,

$$v_n(x) = \inf_{y \in [0, x_1]} \left\{ \overline{u}(y) + \frac{n^2}{2}(x - y)^2 \right\}. \tag{3.38}$$

It can be proved, as in Lemma 5.1 of Fleming and Soner [28] , that v_n is semiconcave and that the inequality

$$0 \leq \overline{u}(x) - v_n(x) \leq 2\frac{k_0^2}{n^2}$$

holds for all $x \in [0, x_1]$ and so $v_n \to \overline{u}$ uniformly. We have that if $x + h \leq x_1$, then $v_n(x+h) - v_n(x) \leq k_0 h$ for $h \leq x_1 - x$. In effect, take $y_0 \in [0, x_1]$ such that $v_n(x) = \overline{u}(y_0) + \frac{n^2}{2}(x - y_0)^2$, we have

$$\begin{aligned} v_n(x+h) - v_n(x) &\leq \left(\overline{u}(y_0 + h) + \tfrac{n^2}{2}(x - y_0)^2\right) - \left(\overline{u}(y_0) + \tfrac{n^2}{2}(x - y_0)^2\right) \\ &= \overline{u}(y_0 + h) - \overline{u}(y_0) \\ &\leq k_0 h. \end{aligned}$$

Since v_n is semiconcave, the set

$$A_n = \left\{x \in [0, x_1] \text{ such that } v_n'(x) \text{ and } v_n''(x) \text{ exist}\right\}$$

has full measure. Therefore $A = \bigcap_{n=1}^{\infty} A_n$ has also full measure. We want to prove that

$$v_n''(x) \leq \frac{8(c+\beta)\overline{u}(x_1)}{\sigma^2 x_0^2} \text{ in } [x_0, x_1] \cap A. \tag{3.39}$$

Take $\bar{x} \in [x_0, x_1] \cap A$ and consider $\bar{y}_n \in [0, x_1]$ such that

$$v_n(\bar{x}) = \overline{u}(\bar{y}_n) + \frac{n^2}{2}(\bar{x} - \bar{y}_n)^2. \tag{3.40}$$

It can be proved that

$$\frac{x_0}{2} \leq \bar{y}_n \leq \bar{x} \text{ and } \bar{x} - \bar{y}_n \leq 2\frac{k_0}{n^2}. \tag{3.41}$$

By (3.38), we have

$$v_n(\bar{x}+h) = \inf_{y \in [0, x_1]} \left\{ \overline{u}(y) + \frac{n^2}{2}(\bar{x} + h - y)^2 \right\} \underset{y = \bar{y}_n + h}{\leq} \overline{u}(\bar{y}_n + h) + \frac{n^2}{2}(\bar{x} - \bar{y}_n)^2$$

so we obtain from (3.40) that

$$\begin{aligned} &\liminf_{h \to 0} \frac{\overline{u}(\bar{y}_n + h) - \overline{u}(\bar{y}_n) - h v_n'(\bar{x}) - \frac{h^2 v_n''(\bar{x})}{2}}{h^2} \\ &\geq \liminf_{h \to 0} \frac{v_n(\bar{x}+h) - v_n(\bar{x}) - h v_n'(\bar{x}) - \frac{h^2 v_n''(\bar{x})}{2}}{h^2} = 0. \end{aligned}$$

Then, we have that $\left(v_n'(\bar{x}), v_n''(\bar{x})\right) \in D_2^- \overline{u}(\bar{y}_n)$. Since $\overline{u}$ is a viscosity supersolution of (2.42) at $\bar{y}_n$, we obtain from Definition 3.6 that for $\gamma \in \Gamma - \{0\}$,

$$\tilde{\mathcal{L}}_\gamma(\overline{u}, v_n'(\bar{x}), v_n''(\bar{x}))(\bar{y}_n) \leq 0. \tag{3.42}$$

If $v_n''(\bar{x}) \leq 0$, the inequality (3.39) holds and if $v_n''(\bar{x}) > 0$, from (3.41) and (3.42), we get that

$$\frac{\sigma^2\gamma^2 x_0^2}{8} v_n''(\bar{x}) \le \frac{\sigma^2\gamma^2\bar{y}^2}{2} v_n''(\bar{x}) \le (c+\beta)\overline{u}(\bar{y}) \le (c+\beta)\overline{u}(x_1)$$

and so we have (3.39). □

Remark 3.5. With a simpler and similar proof of Proposition 3.3 it is possible to prove that the optimal survival probability function with investments defined in (2.28) is a viscosity solution of (2.30). Moreover, taking $c = 0$ in Proposition 3.4, we get that the optimal survival probability function with investments is semiconcave.

3.9 Comments and References

The notion of viscosity solution was introduced by Crandall and Lions [21] for first-order Hamilton–Jacobi equations and by Lions [42] for second-order partial differential equations. It has become a standard tool for control optimization problems; see for instance Fleming and Soner [28] and Bardi and Capuzzo–Dolcetta [13]. Soner [59] and Sayah [53, 54] generalized the notion of viscosity solutions to first-order integrodifferential equations. Let us point out that the notion of viscosity solutions is very useful in the absence of a priori regularity and that sometimes it can be proved, as a second step, that the value function of the problem is indeed smooth. So, many optimization problems can be tackled with minimal assumptions. In the case of mathematical finance, the concept of viscosity solutions is extensively used and there are many articles and books on this subject; we can mention, for instance, Soner's lecture note in [60] and the books of Pham [50] and Touzi [64].

Note that equations (2.30) and (2.42) involve the second-order operator $\sup_{\gamma\in\Gamma}\mathcal{L}_\gamma$; the ellipticity of this operator degenerates at $x = 0$ and could also degenerate at some points $x > 0$ (when the supremum is reached at $\gamma = 0$); so the viscosity solutions of $\sup_{\gamma\in\Gamma}\mathcal{L}_\gamma = 0$ could be a priori non-smooth.

It can be seen that the optimal value functions on the limit diffusion setting are also viscosity solutions of the corresponding HJB equations (2.49) and (2.50), but they turn to be twice continuously differential functions and so they are classical solutions (see for instance [4, 5, 55] and [35]). These equations do not involve second-order degenerate operators since both $\mathcal{L}_D$ and $\tilde{\mathcal{L}}_D$ have positive ellipticity.

Chapter 4
Characterization of Value Functions

This chapter is devoted to characterize the optimal value functions among the viscosity solutions of the corresponding HJB equations in the classical risk model. We consider the bare case presented in (1.6) and (1.10), the case with reinsurance presented in (2.9) and (2.16), and the case with investment presented in (2.28) and (2.33).

The optimal survival probability functions are characterized as the unique nondecreasing viscosity solution of the HJB equation with limit one at infinity (which is the natural boundary condition). This result is also a verification result: if the survival probability function of an admissible strategy is a viscosity solution of the corresponding HJB equation with limit one at infinity, then the admissible strategy and its survival probability function are optimal.

In the problems of optimal dividend payments, we first prove a uniqueness result for viscosity solutions of the HJB equation associated to the optimal dividend problem with boundary condition at zero and linear growth at infinity. The issue here is that the boundary condition at zero for the optimal value function is not known a priori, so this result is not enough to characterize it. We solve this problem by proving that the optimal value function is the smallest viscosity supersolution of the associated HJB equation with linear growth at infinity. We also obtain a verification result: if the value function of an admissible strategy is a viscosity supersolution of the corresponding HJB equation, then the admissible strategy and its value function are optimal.

4.1 Survival Probability

We consider here the simplest case of survival probability function without control. As we pointed out in Sect. 3.1, there is no hope to find a smooth solution of equation (1.13) in the case that the claim-size distribution function is not continuous. We will show in this section that the survival probability function can

P. Azcue and N. Muler, *Stochastic Optimization in Insurance: A Dynamic Programming Approach*, SpringerBriefs in Quantitative Finance, DOI 10.1007/978-1-4939-0995-7_4,

be characterized either as the unique viscosity solution of (1.13) with limit one at infinity or as the unique viscosity solution of (1.13) with boundary condition at zero given by Remark 1.3. For instance, this result allows us to prove that the non-smooth function u defined in (3.3) is indeed the survival probability function in the case that all the claims have size 1.

In this section we also see that the notions of almost-everywhere and viscosity solutions of equation (1.13) coincide.

Proposition 4.1. *There exists a unique almost-everywhere solution W of (1.13) in $\mathbf{R}_+$ with $W(0) = 1$. The function W is nondecreasing and Lipschitz.*

Proof. First we prove that given $x_0 \geq 0$, $a > 0$, and a nondecreasing positive function $f : [0, x_0] \to \mathbf{R}$, there exists a unique almost-everywhere solution W of (1.13) in (x_0, ∞) which satisfies $W = f$ in $[0, x_0]$, and then we prove that W is nondecreasing in (x_0, ∞).

We consider the set $C = \{g : [x_0, \infty) \to \mathbf{R}$ continuous with $g(x_0) = f(x_0)\}$ and we define the operator $T : C \to C$ as

$$T(g)(x) = f(x_0) + \int_{x_0}^{x} \frac{\beta g(s) - \beta \int_0^{s-x_0} g(s-\alpha)dF(\alpha) - \beta \int_{s-x_0}^{s} f(s-\alpha)dF(\alpha)}{p} ds.$$

Taking $x \in [x_0, x_0 + h]$ with $h = p/(4\beta)$, we have

$$\begin{aligned} |T(g_1)(x) - T(g_2)(x)| &\leq \int_{x_0}^{x} \frac{\beta \int_0^{s-x_0} |g_1(s-\alpha) - g_2(s-\alpha)| dF(\alpha) + \beta |g_1(s) - g_2(s)|}{p} ds \\ &\leq \int_{x_0}^{x} \frac{\beta \int_0^{s-x_0} dF(\alpha) + \beta}{p} ds \max_{s \in [x_0, x_0+h]} |g_1(s) - g_2(s)| \\ &\leq \frac{1}{2} \max_{s \in [x_0, x_0+h]} |g_1(s) - g_2(s)| \end{aligned}$$

and so T is a contraction of modulus $1/2$ with respect to the distance $d(g_1, g_2) = \max_{s \in [x_0, x_0+h]} |g_1(s) - g_2(s)|$. Since h does not depend neither on x_0 nor on the function f, there exists a unique $W \in C$ satisfying $T(W) = W$ in $[x_0, \infty)$. Let us define for $x \geq x_0$ the function

$$\hat{W}(x) = \frac{\beta W(x) - \beta \int_0^{x-x_0} W(x-\alpha)dF(\alpha) - \beta \int_{x-x_0}^{x} f(x-\alpha)dF(\alpha)}{p}.$$

We have that W is an absolutely continuous function with $W'(x) = \hat{W}(x)$ at the points where W is differentiable. We also have that W is locally Lipschitz since

$$\frac{\beta W(x) - \beta \max_{s \in [x_0, x]} W(s) - \beta \max_{s \in [0, x_0]} f(s)}{p} \leq \hat{W}(x) \leq \frac{\beta}{p} W(x).$$

Take $u : [x_0, \infty) \to \mathbf{R}$ continuous and nondecreasing with $u(x_0) = W(x_0) > 0$. Since f is nondecreasing, we can see that for any $x \geq x_0$ we obtain

$$\hat{u}(x) = \frac{(c+\beta)u(x)-\beta\int_0^{x-x_0} u(x-\alpha)dF(\alpha)-\beta\int_{x-x_0}^{x} f(x-\alpha)dF(\alpha)}{p} \geq 0,$$

and so $T^{(n)}(u)$ is nondecreasing for all $n \geq 1$ and so, by uniqueness of the fixed point, we have the result. □

Proposition 4.2. *The notions of almost-everywhere solutions and viscosity solutions of (1.13) in* $\mathbf{R}_+$ *are equivalent.*

Proof. By definition, all the viscosity solutions are almost-everywhere solutions. Take an almost-everywhere solution U of (1.13) in $\mathbf{R}_+$, then $U(x)/U(0)$ is an almost-everywhere solution of (1.13) with boundary condition one at zero; by Proposition 4.1, we have that $U(x) = U(0)W(x)$. Let us show that W and so U are viscosity solutions of (1.13). Since F is nondecreasing and right continuous, we have that

$$W'(x^+) = \frac{\beta W(x)-\beta\mathcal{I}(W)(x)}{p} \leq \frac{\beta W(x)-\beta\mathcal{I}(W)(x^-)}{p} = W'(x^-).$$

The function W is differentiable at the continuity points of F and so it is a viscosity solution at these points. At any discontinuity point x of F, the set of sub-differentials is empty and the set of super-differentials is the interval $[W'(x^+), W'(x^-)]$, so W is a viscosity solution of (1.13) at x. □

Theorem 4.1. *The survival probability function defined in (1.6) can be characterized as the unique viscosity solution of (1.13) with boundary condition one at infinity.*

Proof. By Propositions 4.1 and 4.2, all the viscosity solutions of equation (3.3) are multiples of W. By Remark 3.3, we have the result. □

Remark 4.1. Since $\delta(0) = \eta/(1+\eta)$, then the survival probability function δ can also be characterized as the unique viscosity solution of (1.13) with boundary condition $\eta/(1+\eta)$ at zero.

Remark 4.2. The function W defined in Proposition 4.1 satisfies $\lim_{x\to\infty} W(x) < \infty$, and $\delta(x)$ can be written as $W(x)/\lim_{x\to\infty} W(x)$.

4.2 Optimal Dividends

We first address in this section the issue of uniqueness of viscosity solutions for the HJB equation of the dividend problem (1.21) with linear growth at infinity with slope one and a boundary condition at zero. We prove a comparison principle, i.e., given $\underline{u}$ a viscosity subsolution and $\overline{u}$ a viscosity supersolution satisfying the same boundary conditions, then $\underline{u} \leq \overline{u}$. Once we have this result, and since viscosity solutions are both viscosity supersolutions and subsolutions, the uniqueness result holds.

By Proposition 1.2, the optimal dividend value function defined in (1.10) has linear growth with slope one at infinity but does not have a natural boundary condition at zero. So the uniqueness result for viscosity solutions with boundary condition at zero is not a verification theorem (as it is in the case of the survival probability function) because there are infinitely many viscosity solutions of (1.21) with slope one at infinity: considering, for instance, the functions $u(x) = k + x$ for any $k \geq p/c$, we have that $u'(x) = 1$ and

$$\tilde{\mathcal{L}}(k + x) \leq p - c\,(x + k) \leq p - ck \leq 0$$

for all $x \geq 0$, where the operator $\tilde{\mathcal{L}}$ is defined in (1.22).

In order to deal with this issue, we obtain the following characterization result: the value function given in (1.10) is the smallest supersolution (and since it is a viscosity solution, it is the smallest viscosity solution as well). Finally, we deduce a verification result: any value function of an admissible strategy that is a viscosity supersolution of (1.21) is the optimal value function.

Let us define the natural growth condition at infinity for the optimal value function.

Definition 4.1. A continuous function u has *growth condition A.1* if there exists a constant $k > 0$ such that $u(x) \leq x + k$ for all $x \in \mathbf{R}_+$.

The following result is called the comparison principle for (1.21).

Proposition 4.3. *Let us assume that for all $x > 0$, $\underline{u}(x)$ is a subsolution of (1.21) and $\overline{u}(x)$ is a supersolution of (1.21), both nondecreasing and nonnegative with growth condition A.1. If $\underline{u}(0) = \overline{u}(0)$, then $\underline{u} \leq \overline{u}$ in $\mathbf{R}_+$.*

Proof. By Definition 3.3, the function $\overline{u}$ is locally Lipschitz and satisfies $\overline{u}(x) - \overline{u}(y) \geq x - y$ for $x > y$. Assume that there is a point x_0 such that $\underline{u}(x_0) - \overline{u}(x_0) > 0$. We consider $\overline{u}^s(x) = s\overline{u}(x)$ with $s > 1$. It is easy to see that $\overline{u}^s(x)$ is also a supersolution with $\overline{u}^s(0) \geq \underline{u}(0)$ and also satisfies $\overline{u}^s(x) - \overline{u}^s(y) \geq s\,(x - y)$ for $x > y$. We can choose $s > 1$ such that $\underline{u}(x_0) - \overline{u}^s(x_0) > 0$, from the growth condition A.1 we obtain

$$\underline{u}(x) - \overline{u}^s(x) \leq k + (1 - s)x,$$

and so we have

$$\underline{u}(x) - \overline{u}^s(x) \leq 0 \text{ for } x \geq \frac{k}{s-1}. \tag{4.1}$$

Let us define

$$M = \sup_{x \geq 0} (\underline{u}(x) - \overline{u}^s(x)), \tag{4.2}$$

then from (4.1) we obtain

$$0 < \underline{u}(x_0) - \overline{u}^s(x_0) \leq M = \max_{x \in [0,b]} (\underline{u}(x) - \overline{u}^s(x)), \tag{4.3}$$

where $b = k/(s-1)$. Call $x^* = \arg\max_{x\in[0,b]} (\underline{u}(x) - \overline{u}^s(x))$. Since $\underline{u}(x)$ and $\overline{u}^s(x)$ are locally Lipschitz, there exists a constant $m > 0$ such that

$$\frac{\underline{u}(x_1) - \underline{u}(x_2)}{x_1 - x_2} \leq m, \quad \frac{\overline{u}^s(x_1) - \overline{u}^s(x_2)}{x_1 - x_2} \leq m \tag{4.4}$$

for $0 \leq x_2 \leq x_1 \leq b$. Let us consider

$$A = \{(x, y) : 0 \leq y \leq b, 0 \leq x \leq y\},$$

and for any $\lambda > 0$ the functions

$$\Phi^\lambda(x, y) = \frac{\lambda}{2}(x - y)^2 + \frac{2m}{\lambda^2(y - x) + \lambda} \tag{4.5}$$

and

$$\Sigma_\lambda(x, y) = \underline{u}(x) - \overline{u}^s(y) - \Phi^\lambda(x, y). \tag{4.6}$$

Calling

$$M_\lambda = \max_A \Sigma_\lambda \tag{4.7}$$

and $(x_\lambda, y_\lambda) = \arg\max_A \Sigma_\lambda$,we obtain that

$$M_\lambda \geq \Sigma_\lambda(x^*, x^*) = M - 2m/\lambda$$

and so from (4.3) we get that $M_\lambda > 0$ for $\lambda \geq 4m/M$ and

$$\liminf_{\lambda\to\infty} M_\lambda \geq M. \tag{4.8}$$

Since $(x_\lambda, y_\lambda) \in A$, we have that

$$y_\lambda \geq x_\lambda. \tag{4.9}$$

First we show that there exists λ_0 large enough such that for any $\lambda \geq \lambda_0$ we have that $(x_\lambda, y_\lambda) \notin \partial A$. Since $\underline{u}$ is an increasing function, we have from (4.1) and (4.6) that

$$\Sigma_\lambda(x, b) \leq \underline{u}(x) - \overline{u}^s(b) \leq \underline{u}(b) - \overline{u}^s(b) < 0. \tag{4.10}$$

Also we have from (4.4) and (4.6) that for $x > 0$,

$$\limsup_{h\to 0+} \frac{\Sigma_\lambda(x, x) - \Sigma_\lambda(x - h, x)}{h} \leq -m < 0. \tag{4.11}$$

Finally, we have on one hand that

$$\Sigma_\lambda\,(0,0) = \underline{u}(0) - \overline{u}^s(0) - \frac{2m}{\lambda} < 0, \tag{4.12}$$

and so there exists $\varepsilon > 0$ such that $\Sigma_\lambda\,(0,y) < 0$ for all $y \in [0,\varepsilon]$, and on the other hand we have from (4.4) and (4.6) that for $y \geq \varepsilon$

$$\limsup_{h\to 0+} \frac{\Sigma_\lambda\,(0,y) - \Sigma_\lambda\,(h,y)}{h} \leq 2m - \lambda y < 0 \tag{4.13}$$

for $\lambda > 2m/\varepsilon$. Taking $\lambda_0 = \max\{2m/\varepsilon, 4m/M\}$, since $M_\lambda > 0$ for $\lambda \geq \lambda_0$, and combining (4.10)–(4.13), we have that $(x_\lambda, y_\lambda) \notin \partial A$ for $\lambda \geq \lambda_0$. Since Σ_λ attained a local maximum at (x_λ, y_λ), we have

$$\underline{u}(x) - \Phi^\lambda\,(x, y_\lambda) \leq \underline{u}(x_\lambda) - \Phi^\lambda\,(x_\lambda, y_\lambda) \tag{4.14}$$

for x near x_λ. Then,

$$\limsup_{x\to x_\lambda} \frac{\underline{u}(x) - \underline{u}(x_\lambda)}{x - x_\lambda} \leq \Phi^\lambda_x\,(x_\lambda, y_\lambda)$$

and so

$$\Phi^\lambda_x(x_\lambda, y_\lambda) \in D_2^+\,(\underline{u})\,(x_\lambda). \tag{4.15}$$

Analogously, we get

$$-\,\Phi^\lambda_y(x_\lambda, y_\lambda) \in D_2^-\,(\overline{u}^s)\,(y_\lambda). \tag{4.16}$$

Therefore, if we call

$$\tilde{\mathcal{L}}(u,d)(x) = pd - (c+\beta)u(x) + \beta\mathcal{I}(u)(x),$$

we have the following inequalities:

$$\max\{1 - \Phi^\lambda_x(x_\lambda, y_\lambda), \tilde{\mathcal{L}}(\underline{u}, \Phi^\lambda_x(x_\lambda, y_\lambda))(x_\lambda)\} \geq 0, \tag{4.17}$$

$$\max\{1 + \Phi^\lambda_y(x_\lambda, y_\lambda), \tilde{\mathcal{L}}(\overline{u}^s, -\Phi^\lambda_y(x_\lambda, y_\lambda))(y_\lambda)\} \leq 0. \tag{4.18}$$

Since $1 + \Phi^\lambda_y(x_\lambda, y_\lambda) \leq 1 - s < 0$ and $\Phi^\lambda_x(x_\lambda, y_\lambda) = -\Phi^\lambda_y(x_\lambda, y_\lambda)$ we have from (4.17) that

$$\tilde{\mathcal{L}}(\underline{u}, \Phi^\lambda_x(x_\lambda, y_\lambda))(x_\lambda) \geq 0. \tag{4.19}$$

Therefore, from (4.18) and (4.19), we get

$$(c+\beta)\left(\underline{u}(x_\lambda)-\overline{u}^s(y_\lambda)\right) \le \beta\left(\int_0^{x_\lambda}\underline{u}(x_\lambda-\alpha)dF(\alpha)-\int_0^{y_\lambda}\overline{u}^s(y_\lambda-\alpha)dF(\alpha)\right). \tag{4.20}$$

Using the inequality

$$\Sigma_\lambda\left(x_\lambda,x_\lambda\right)+\Sigma_\lambda\left(y_\lambda,y_\lambda\right)\le 2\Sigma_\lambda\left(x_\lambda,y_\lambda\right),$$

we obtain

$$\lambda\left(x_\lambda-y_\lambda\right)^2\le \underline{u}(x_\lambda)-\underline{u}(y_\lambda)+\overline{u}^s(x_\lambda)-\overline{u}^s(y_\lambda)+4m(y_\lambda-x_\lambda);$$

then we have, from (4.4), that

$$\lambda\left(x_\lambda-y_\lambda\right)^2\le 6m\left|x_\lambda-y_\lambda\right|. \tag{4.21}$$

We can find a sequence $\lambda_n\to\infty$ such that $(x_{\lambda_n},y_{\lambda_n})\to(\overline{x},\overline{y})\in A$. From (4.21), we get that $|x_{\lambda_n}-y_{\lambda_n}|\le 6m/\lambda_n$ and this gives $\overline{x}=\overline{y}$. Using that $y_{\lambda_n}\ge x_{\lambda_n}$ for all n, we obtain from (4.20)

$$(c+\beta)\left(\underline{u}(\overline{x})-\overline{u}^s(\overline{x})\right)\le\beta\int_0^C\left(\underline{u}(\overline{x}-\alpha)-\overline{u}^s(\overline{x}-\alpha)\right)dF(\alpha)\le\beta M, \tag{4.22}$$

where C can be equal to either $\overline{x}$ or $\overline{x}^-$. From (4.21), we get that $\lim_{n\to\infty}\lambda_n\left(x_{\lambda_n}-y_{\lambda_n}\right)^2=0$; hence from (4.8) and (4.22) we obtain

$$M\le\liminf_{\lambda\to\infty}M_\lambda\le\lim_{n\to\infty}M_{\lambda_n}=\underline{u}(\overline{x})-\overline{u}^s(\overline{x})\le\frac{\beta}{c+\beta}M, \tag{4.23}$$

which is a contradiction. □

As we have pointed out before, the following uniqueness result for viscosity solutions is a direct consequence of the previous proposition.

Corollary 4.1. *There is at most one viscosity solution of (1.21) with boundary condition $u(0)=u_0$ among all the functions that satisfy growth condition A.1.*

In order to obtain the characterization result, we have to prove the following technical lemma.

Lemma 4.1. *Fix $x_0>0$ and let $\overline{u}$ be a nonnegative supersolution of (1.21) satisfying the growth condition A.1. We can find a sequence of positive functions $\overline{u}_n:R_+\to R$ such that:*

(a) $\overline{u}_n$ is continuously differentiable.
(b) $\overline{u}_n$ satisfies the growth condition A.1.

(c) $1 \le \overline{u}_n'(x) \le (c+\beta)\,\overline{u}_n(x)/p$.
(d) $u_n \searrow \overline{u}$ *uniformly on compact sets and* $\overline{u}_n'(x)$ *converges to* $\overline{u}'(x)$ *a.e.*
(e) There exists a sequence c_n *with* $\lim_{n\to\infty} c_n = 0$ *such that* $\sup_{x\in[0,x_0]} \tilde{\mathcal{L}}(\overline{u}_n)(x) \le c_n$.

Proof. Since $\overline{u}$ is an absolutely continuous supersolution of (1.21), we have that

$$\overline{u}'(x) \le \frac{c+\beta}{p}\overline{u}(x) - \frac{\beta}{p}\int_0^\infty \overline{u}(x-\alpha)dF(\alpha) \le \frac{c+\beta}{p}\overline{u}(x) \text{ a.e.} \tag{4.24}$$

and $\overline{u}'(x) \ge 1$ a.e.; this implies that $\overline{u}(x)$ is positive and increasing for all $x \ge 0$.

Let $\phi(x)$ be a nonnegative continuously differentiable function with support included in $(0,1)$ such that $\int_0^1 \phi(x) = 1$; we define $\overline{u}_n : \mathbf{R}_+ \to R$ as the convolution

$$\overline{u}_n(x) = \int_{-\infty}^{\infty} \overline{u}(x+s)n\phi(ns)ds. \tag{4.25}$$

By definition, $\overline{u}_n(x)$ is a weighted average of values of $\overline{u}(y)$ for $y \in [x, x+1/n]$. We use this non-centered average in order to take advantage of the right continuity of F. We have that $\overline{u}_n \ge \overline{u}$ and since $\overline{u}$ is absolutely continuous in $\mathbf{R}_+$ and satisfies growth condition A.1, (a), (b), and (d) follow by standard techniques; see for instance [67]. From (4.24) and using that $\overline{u}'(x) \ge 1$ a.e., we conclude (c). Let us define for $x \in [0, x_0]$ the function

$$\xi_n(x) = \sup\left\{\overline{u}'(y) \text{ for } y \in [x, x+1/n] \text{ with } \overline{u} \text{ differentiable at } y\right\}, \tag{4.26}$$

we have from (4.24) and (4.25) that

$$\overline{u}_n'(x) \le \xi_n(x) \le \frac{c+\beta}{p}\overline{u}(x_0). \tag{4.27}$$

From (4.26), there exists $y_n \in [x, x+1/n]$ such that

$$\overline{u}'(y_n) \ge \xi_n(x) - \frac{1}{n}. \tag{4.28}$$

We conclude from (4.24), (4.27), and (4.28) that, for any $x \in [0, x_0]$,

$$\begin{aligned}\tilde{\mathcal{L}}(\overline{u}_n)(x) &= \tilde{\mathcal{L}}(\overline{u})(y_n) + \tilde{\mathcal{L}}(\overline{u}_n)(x) - \tilde{\mathcal{L}}(\overline{u})(y_n)\\ &\le \tfrac{p}{n} - (c+\beta)(\overline{u}_n(x) - \overline{u}(x)) + (c+\beta)(\overline{u}(y_n) - \overline{u}(x))\\ &\quad +\beta\int_0^{y_n} (\overline{u}_n(x-\alpha) - \overline{u}(y_n-\alpha))\, dF(\alpha)\\ &\le c_n,\end{aligned}$$

where

$$c_n = \left(p + \frac{(c+\beta)^2\overline{u}(x_0)}{p}\right)\frac{1}{n} + \beta \sup_{y\in[0,x_0]} (\overline{u}_n(y) - \overline{u}(y)).$$

□

Proposition 4.4. *The optimal value function defined in (1.10) is the smallest viscosity supersolution of (1.21) satisfying growth condition A.1.*

Proof. Let $\overline{u}$ be a nonnegative supersolution of (1.21) satisfying the growth condition A.1 and let $\overline{L} = (L_t)_{t\geq 0} \in \Pi_x^L$ be any admissible strategy; define X_t as the corresponding controlled risk process starting at x. Since the functions $\overline{u}_n$ defined in Lemma 4.1 are continuously differentiable, we obtain using Proposition 2.13 that

$$\overline{u}_n(X_{t\wedge\tau})e^{-c(t\wedge\tau)} - \overline{u}_n(x) \leq \int_0^{t\wedge\tau} \tilde{\mathcal{L}}(\overline{u}_n)(X_{s}-)e^{-cs}ds - \int_0^{t\wedge\tau} e^{-cs}dL_s + \tilde{M}_{t\wedge\tau}, \tag{4.29}$$

where $\tilde{M}_t$ is a zero-expectation martingale. Here, the functions $\overline{u}_n$ are extended as $\overline{u}_n(x) = 0$ for $x < 0$.

Using that L_t is increasing we get, using the monotone convergence theorem, that

$$\lim_{t\to\infty} E_x\left(\int_0^{t\wedge\tau} e^{-cs}dL_s\right) = E_x\left(\int_0^{\tau} e^{-cs}dL_s\right) = V_{\overline{L}}(x), \tag{4.30}$$

and from Lemma 4.1(c), we have

$$-(c+\beta)\overline{u}_n(x) \leq \tilde{\mathcal{L}}(\overline{u}_n)(x) \leq \beta\overline{u}_n(x).$$

But using Lemma 4.1(b) and the inequality $X_s \leq x + ps$, we get

$$\overline{u}_n(X_s) \leq k_0 + X_s \leq k_0 + x + ps \tag{4.31}$$

for some k_0. So, using the bounded convergence theorem, we obtain

$$\lim_{t\to\infty} E_x\left(\int_0^{t\wedge\tau} \tilde{\mathcal{L}}(\overline{u}_n)(X_{s}-)e^{-cs}ds\right) = E_x\left(\int_0^{\tau} \tilde{\mathcal{L}}(\overline{u}_n)(X_{s}-)e^{-cs}ds\right). \tag{4.32}$$

From (4.29), (4.30), and (4.32), we get

$$\lim_{t\to\infty} E_x\left(\overline{u}_n(X_{t\wedge\tau})e^{-c(t\wedge\tau)}\right) - \overline{u}_n(x) \leq E_x\left(\int_0^{\tau} \tilde{\mathcal{L}}(\overline{u}_n)(X_{s}-)e^{-cs}ds\right) - V_{\overline{L}}(x). \tag{4.33}$$

We show next that

$$\lim_{t\to\infty} E_x\left(\overline{u}_n(X_{t\wedge\tau})e^{-c(t\wedge\tau)}\right) = 0. \tag{4.34}$$

From (4.31), we have

$$\begin{aligned} E_x\left(\overline{u}_n(X_{t\wedge\tau})e^{-c(t\wedge\tau)}\right) &= E_x\left(\overline{u}_n(X_{t\wedge\tau})e^{-c(t\wedge\tau)}I_{\{t<\tau\}}\right) \\ &\leq E_x\left(\overline{u}_n(X_t)e^{-ct}\right) \\ &\leq (k_0 + x + pt)\,e^{-ct}. \end{aligned}$$

Since the last expression goes to 0 as t goes to infinity we have (4.34). Let us prove now that

$$\limsup_{n\to\infty} E_x\left(\int_0^{\tau} \tilde{\mathcal{L}}(\overline{u}_n)(X_{s}-)e^{-cs}ds\right) \leq 0. \tag{4.35}$$

Given any $\varepsilon > 0$, we can find T such that

$$\int_T^{\infty} \tilde{\mathcal{L}}(\overline{u}_n)(X_{s}-)e^{-cs}ds < \frac{\varepsilon}{2} \tag{4.36}$$

Because, by (4.31), growth condition A.1, and Lemma 4.1(b) and (c), there exists a constant k such that

$$\begin{aligned}\tilde{\mathcal{L}}(\overline{u}_n)(X_{s}-) &\leq p\overline{u}_n'(X_{s}-) + \beta\int_0^{\infty} \overline{u}_n(X_{s}- - \alpha)dF(\alpha)\\ &\leq (c+\beta)\,\overline{u}_n(X_{s}-) + \beta\overline{u}_n(X_{s}-)\\ &\leq (c+2\beta)X_{s}- + k\\ &\leq (c+2\beta)(k_0 + x + ps) + k.\end{aligned}$$

Note that for $s \leq T$, we have that $X_{s}- \leq x_0 := k_0 + x + pT$. From Lemma 4.1(e) we can find n_0 large enough such that for any $n \geq n_0$

$$\int_0^{T} \tilde{\mathcal{L}}(\overline{u}_n)(X_{s}-)e^{-cs}ds \leq \frac{c_n}{c} \leq \frac{\varepsilon}{2}$$

and so we get (4.35). Then, from (4.33) and using (4.34) and (4.35), we obtain

$$\overline{u}(x) = \lim_{n\to\infty} \overline{u}_n(x) \geq V_{\overline{L}}(x).$$

So since V is a viscosity solution of (1.21), the result follows. □

From (1.10) and Proposition 4.4, we get the following verification theorem.

Theorem 4.2. *Consider a family of admissible strategies $(\overline{L}^x)_{x\geq 0}$ such that $\overline{L}^x \in \Pi_x^L$ for any initial surplus $x \geq 0$. If the function $V_{\overline{L}^x}(x)$ is a viscosity supersolution of (1.21), then $V_{\overline{L}^x}(x)$ is the optimal value function.*

Remark 4.3. The Perron's method (see for instance Ishii [37]) gives existence and uniqueness of viscosity solutions provided that two condition holds: the comparison principle we already proved in Proposition 4.3 and the existence of a super- and subsolution with the same boundary condition at zero and boundary condition A.1 at infinity. Given any boundary condition at zero $a \geq V(0)$, it is easy to check that $\overline{u}(x) = V(x) - V(0) + a$ and $\underline{u}(x) = x + a$ are viscosity super- and subsolutions of (1.21), respectively, and satisfy the same boundary conditions. Therefore, there exists a unique viscosity solution of the equation with boundary condition a at zero and A.1 at infinity for any $a \geq V(0)$.

Remark 4.4. On the other hand, by the previous Corollary, there is no viscosity solution of (1.21) if the boundary condition at zero is less than $V(0)$.

The previous two remarks also hold for the dividend problems with reinsurance or investment controls.

Remark 4.5. The simple example introduced in Sect. 3.4 corresponds to the optimal value function (1.10) and the HJB equation (1.21) in the limit case with $\beta \to 0$.

4.3 Optimal Survival Probability with Reinsurance

In this section we characterize the optimal survival probability function with reinsurance control defined in (2.9) as the unique viscosity solution of the HJB equation (2.13) with limit one at infinity. As in Sect. 4.1, there is no hope to obtain in general smooth solutions if the claim-size distribution is not continuous.

The usual way to prove uniqueness of viscosity solutions is using a comparison-principle argument, as we did in Proposition 4.3, but in this problem we were not able to obtain a contradiction similar to the one in (4.23). Instead, we prove the uniqueness viscosity result by showing that the optimal survival probability function is both the smallest viscosity supersolution and the largest viscosity subsolution, with limit one at infinity.

The use of a fixed-point argument analogous to Proposition 4.1 to prove uniqueness for general claim-size distributions is much more involved than the simpler case without reinsurance, so we are not using this approach. However, in the case that the claim-size distribution has bounded density and for some one-parameter families of reinsurance contracts, one can use a fixed-point argument to show uniqueness of smooth solution (see [34, 55, 57]).

In order to prove the uniqueness of viscosity solutions, we use the following three technical lemmas. The proof of the first one is the same as the proof of Lemma 4.1. The second lemma is similar to the first but deals with subsolutions.

Lemma 4.2. *Let $\overline{u}$ be a nondecreasing Lipschitz supersolution of (2.13) satisfying* $\lim_{x\to\infty} \overline{u}(x) = 1$. *We can find a sequence of positive functions $\overline{u}_n : R_+ \to R$ such that:*

(a) *$\overline{u}_n$ is infinitely continuously differentiable and $\overline{u}_n' \leq K$, where K is the Lipschitz constant of $\overline{u}$.*
(b) $\lim_{x\to\infty} \overline{u}_n(x) = 1$.
(c) *$\overline{u}_n \searrow \overline{u}$ uniformly in $\mathbf{R}_+$ and $\overline{u}_n'(x)$ converges to $\overline{u}'(x)$ a.e.*
(d) *There exists a sequence $c_n > 0$ with $\lim\limits_{n\to\infty} c_n = 0$ such that*

$$\sup_{R\in\mathcal{R}, x\geq 0} \mathcal{L}_R(\overline{u}_n)(x) \leq c_n.$$

Lemma 4.3. *Let $\underline{u}$ be a nondecreasing Lipschitz subsolution of (2.13) satisfying* $\lim_{x\to\infty} \underline{u}(x) = 1$*. We can find a sequence of positive functions $\underline{u}_n : R_+ \to R$ such that:*

(a) $\underline{u}_n$ is infinitely continuously differentiable and $\underline{u}_n' \leq K$, where K is the Lipschitz constant of $\underline{u}$.
(b) $\lim_{x\to\infty} \underline{u}_n(x) = 1$.
(c) $\underline{u}_n \nearrow \underline{u}$ uniformly in $\mathbf{R}_+$ and $\underline{u}_n'(x)$ converges to $\underline{u}'(x)$ a.e.
(d) There exists a sequence $c_n > 0$ with $\lim_{n\to\infty} c_n = 0$ such that

$$\sup_{R\in\mathcal{R}, x\in\mathbf{R}_+} \mathcal{L}_R(\underline{u}_n)(x) \geq -c_n.$$

Proof. The argument is similar to the one in the proof of Lemma 4.1, but taking the convolution

$$\underline{u}_n(x) = \int_{-\infty}^{\infty} \underline{u}(x+s) n\phi(ns)ds \tag{4.37}$$

with a nonnegative continuously differentiable function ϕ with support included in $(-1, 0)$ such that $\int_{-1}^{0} \phi(x) = 1$. □

Lemma 4.4. *Given any twice continuously differentiable function u and any $\varepsilon > 0$ there exists a stationary reinsurance strategy ρ, as defined in Definition 2.2, such that $\sup_{R\in\mathcal{R}} \mathcal{L}_R(x) - \mathcal{L}_{\rho^x}(x) < \varepsilon$ for all $x \geq 0$.*

Proof. Define $H(x) = \sup_{R\in\mathcal{R}} \mathcal{L}_R(x)$ and take $\varepsilon > 0$. From Lemma 3.2, we can find, using (3.27) and (3.28), a sequence $0 = x_0 < x_1 < x_2 < \cdots$ such that $\lim_{n\to\infty} x_n = \infty$ satisfying

$$H(x) - H(x_n) < \varepsilon/3 \tag{4.38}$$

and

$$\mathcal{L}_R(x) - \mathcal{L}_R(x_n) < \varepsilon/3 \tag{4.39}$$

for any $x \in [x_n, x_{n+1})$ and any $R \in \mathcal{R}$.

Consider a retained loss function $R^n \in \mathcal{R}$ such that

$$H(x_n) - \mathcal{L}_{R^n}(x_n) < \varepsilon/3, \tag{4.40}$$

and define

$$\rho^x = \sum_{n=0}^{\infty} R^n I_{[x_n, x_{n+1})}(x),$$

then we have that ρ is a stationary reinsurance control and

$$H(x) - \mathcal{L}_{\rho^x}(x) < \varepsilon$$

for all $x > 0$. In effect, given $x \in [x_n, x_{n+1})$, we have, by (4.38)–(4.40), that

$$\begin{aligned} &H(x) - \mathcal{L}_{\rho^x}(x) \\ &= (H(x) - H(x_n)) + (H(x_n) - \mathcal{L}_{R^n}(x_n)) + \left(\mathcal{L}_{R^n}(x_n) - \mathcal{L}_{\rho^x}(x)\right) \\ &< \varepsilon. \end{aligned}$$

□

Theorem 4.3. *The optimal survival probability function can be characterized as the unique nondecreasing viscosity solution of (2.13) with limit one at infinity.*

Proof. Let us prove first that the optimal value function is smaller or equal than any supersolution. Take $\overline{u}$ a nondecreasing supersolution of (2.13) in $(0, \infty)$ with $\lim_{x\to\infty} \overline{u}(x) = 1$. Consider $x \geq 0$ and any admissible strategy $R = (R_t)_{t\geq 0} \in \Pi_x^R$. Define X_t as the corresponding controlled risk process with initial surplus x and let τ be its ruin time. For any $M > x$, we also define the stopping times

$$\tau_M = \inf\left\{t \geq 0 : X_t^R \geq M\right\}.$$

Considering the functions $\overline{u}_n$ defined in Lemma 4.2, we extend the definition of $\overline{u}_n$ as $\overline{u}_n = 0$ in $(-\infty, 0)$. Using Proposition 2.12 and Lemma 4.2(d), we obtain

$$\begin{aligned} &\overline{u}_n(X_{\tau_M\wedge\tau\wedge t}) - \overline{u}_n(x) \\ &= \int_0^{\tau_M\wedge\tau\wedge t} \mathcal{L}_{R_s}(\overline{u}_n)(X_{s}-)ds + M_{\tau_M\wedge\tau\wedge t} \\ &\leq c_n\left(\tau_M \wedge \tau \wedge t\right) + M_{\tau_M\wedge\tau\wedge t}, \end{aligned}$$

where M_t is a zero-expectation martingale. So we get

$$E_x\left(\overline{u}_n(X_{\tau_M\wedge\tau\wedge t}) - \overline{u}_n(x)\right) \leq c_n t.$$

Taking the limit as $n \to \infty$ (with t fixed) we obtain

$$\limsup_{n\to\infty} E_x\left((\overline{u}_n(X_{\tau_M\wedge\tau\wedge t})) - \overline{u}_n(x)\right) \leq 0$$

and so, since $\lim_{n\to\infty} \overline{u}_n(x) = \overline{u}(x)$ and $\overline{u} \leq \overline{u}_n$, we have

$$E_x\left(\overline{u}(X_{\tau_M\wedge\tau\wedge t})\right) - \overline{u}(x) \leq 0.$$

So taking $t \to \infty$,

$$E_x\left(\overline{u}(X_{\tau_M\wedge\tau})\right) - \overline{u}(x) = \overline{u}(M)P(\tau_M < \tau) - \overline{u}(x) \leq 0.$$

Now taking $M \to \infty$, and since $\lim_{M\to\infty} \overline{u}(M) = 1$ and $\lim_{M\to\infty} P(\tau_M < \tau) = \delta^{\overline{R}}(x)$, we get $\delta^{\overline{R}}(x) \le \overline{u}(x)$ for all $\overline{R} \in \Pi_x^R$ and so

$$\delta(x) \le \overline{u}(x). \tag{4.41}$$

Finally, let us prove that the optimal value function is greater or equal than any subsolution. Take a nondecreasing subsolution $\underline{u}$ of (2.13) in $(0,\infty)$ with $\lim_{x\to\infty} \underline{u}(x) = 1$. Considering the functions $\underline{u}_n$ defined in Lemma 4.3, we extend the definition of $\underline{u}_n$ as $\underline{u}_n = 0$ in $(-\infty, 0)$. By Lemma 4.4, there exists a stationary reinsurance control ρ_n such that

$$\sup_{R\in\mathcal{R}} \mathcal{L}_R(\underline{u}_n)(y) - \mathcal{L}_{\rho_n^y}(\underline{u}_n)(y) \le 1/n$$

for all $y \ge 0$ and $n \ge 1$. Let us consider the controlled process $(X_t^n)_{t\ge0}$ with initial surplus x and admissible reinsurance strategy $\overline{R}^n = \left(R_t^n\right)_{t\ge0} = \left(\rho_n^{X_t-}\right)_{t\ge0}$ associated to this stationary reinsurance control. Let us define the corresponding ruin time τ^n and, for any $M > 0$, the stopping time

$$\tau_M^n = \inf\{t \ge 0 : X_t^n \ge M\}.$$

Using Proposition 2.12, we obtain for each n that

$$\begin{aligned}
&\underline{u}_n(X^n_{\tau_M^n\wedge\tau^n\wedge t}) - \underline{u}_n(x)\\
&= \int_0^{\tau_M^n\wedge\tau^n\wedge t} \mathcal{L}_{R_s^n}(\underline{u}_n)(X_{s-}^n)ds + M_{\tau_M^n\wedge\tau^n\wedge t}\\
&\ge \int_0^{\tau_M^n\wedge\tau^n\wedge t} \left(\sup_{R\in\mathcal{R}}\left(\mathcal{L}_R(\underline{u}_n)(X_{s-}^n)\right) - \tfrac{1}{n}\right) ds + M_{\tau_M^n\wedge\tau^n\wedge t}\\
&\ge \left(-c_n - \tfrac{1}{n}\right)\left(\tau_M^n \wedge \tau^n \wedge t\right) + M_{\tau_M^n\wedge\tau^n\wedge t},
\end{aligned}$$

where M_t is a martingale with zero expectation. So we get

$$E_x\left(\underline{u}_n(X^n_{\tau_M^n\wedge\tau^n\wedge t}) - \underline{u}_n(x)\right) \ge \left(-c_n - \frac{1}{n}\right) E_x\left(\tau_M^{R_n} \wedge \tau^{R_n} \wedge t\right).$$

Taking the limit as $n \to \infty$ with t fixed, we have

$$\liminf_{n\to\infty} E_x\left(\underline{u}_n(X^n_{\tau_M^n\wedge\tau^n\wedge t})\right) \ge \underline{u}(x).$$

Since $\underline{u}_n \le \underline{u}$ we obtain

$$\liminf_{n\to\infty} E_x\left(\left(\underline{u}(X^n_{\tau_M^n\wedge\tau^n\wedge t})\right)\right) \ge \underline{u}(x).$$

For $\varepsilon > 0$, take n_0 large enough so

$$E_x\left(\underline{u}(X^{n_0}_{\tau_M^{n_0}\wedge\tau^{n_0}\wedge t})\right) \ge \underline{u}(x) - \varepsilon.$$

Then,

$$\underline{u}(x) - \varepsilon \leq E_x\left(\underline{u}(X^{n_0}_{\tau^{n_0}_M \wedge \tau^{n_0} \wedge t})\right)$$
$$= \underline{u}(M)P(\tau^{n_0}_M < \tau^{n_0} \wedge t) + E_x\left(\underline{u}(X^{n_0}_{\tau^{n_0}_M \wedge \tau^{n_0} \wedge t})I_{t<\tau^{n_0}_M \wedge \tau^{n_0}}\right).$$

Since $P(\tau^{n_0}_M < \tau^{n_0} \wedge t)$ is nondecreasing on t, $0 \leq \underline{u} \leq 1$, and $\lim_{t\to\infty} P(\tau^{n_0}_M \wedge \tau^{n_0} > t) = 0$, we get, taking $t \to \infty$,

$$\underline{u}(x) - \varepsilon \leq \liminf_{t\to\infty} E_x\left(\underline{u}(X^{n_0}_{\tau^{n_0}_M \wedge \tau^{n_0} \wedge t})\right) = \underline{u}(M)P(\tau^{n_0}_M < \tau^{n_0}).$$

Finally, taking $M \to \infty$, we obtain

$$\underline{u}(x) - \varepsilon \leq \delta^{\overline{R}^{n_0}}(x) \leq \delta(x)$$

and so $\underline{u}(x) \leq \delta(x)$. □

Remark 4.6. From Proposition 4.3, we can see that any nonnegative, bounded nondecreasing viscosity solution of (2.13) can be written as a nonnegative multiple of the optimal survival probability function δ.

4.4 Optimal Dividends and Reinsurance

The results of Proposition 4.3, Corollary 4.1, and Proposition 4.4 also hold for the optimal value function (2.16) and the corresponding HJB equation (2.19). We also have the following verification theorem.

Theorem 4.4. *Consider a family of admissible strategies $(\overline{L}^x, \overline{R}^x)_{x\geq 0}$ such that $(\overline{L}^x, \overline{R}^x) \in \Pi^{L,R}_x$ for any $x \geq 0$. If the function $V_{\overline{L}^x,\overline{R}^x}(x)$ is a viscosity supersolution of (1.21), then $V_{\overline{L}^x,\overline{R}^x}(x)$ is the optimal value function.*

For the proofs of these results, see Propositions 4.2 and 5.1, Corollary 5.2, and Theorem 5.3 in [9].

4.5 Investments and Survival Probability

We study in this section the problem of optimal survival probability with investments defined in (2.28). In order to prove that there exists a unique viscosity solution of (2.30) with limit one at infinity, we use a comparison principle. Equation (2.30) involves the second-order operator $\sup_{\gamma\in\Gamma} \mathcal{L}_\gamma$ whose ellipticity could degenerate. However, the solution is actually smooth as it is proved both in [32] without borrowing constraints and in [10] with $\Gamma = [0, \hat{\gamma}]$. In both works, a fixed-point approach is used.

We use here a comparison-principle result to show uniqueness because it can be generalized to more general cases (see [26]).

Proposition 4.5. *Assume that either $F(x) < 1$ for all $x \geq 0$ or F is absolutely continuous. Consider two positive, nondecreasing, and Lipschitz functions $\underline{u}$ and $\overline{u}$ in $(0, \infty)$. If $\underline{u}$ is a subsolution and $\overline{u}$ is a supersolution of (2.30) in $(0, \infty)$ with* $\lim_{x\to\infty} \overline{u}(x) = \lim_{x\to\infty} \underline{u}(x) = 1$, *then $\underline{u} \leq \overline{u}$ in $\mathbf{R}_+$.*

Proof. Assume that there is a point $x_0 \geq 0$ with $\underline{u}(x_0) - \overline{u}(x_0) > 0$. Take $\varepsilon > 0$ small enough such that $\underline{u}(x_0) - s\overline{u}(x_0) > 0$ for $s \in [1, 1+\varepsilon]$. Let us call $\overline{u}^s = s\overline{u}$. Since $\lim_{x\to\infty} \underline{u}(x) - \overline{u}^s(x) = 1 - s < 0$, we get that there exists $b_s > 0$ such that

$$\underline{u}(x) - \overline{u}^s(x) \leq 0 \text{ for } x \geq b_s. \tag{4.42}$$

Let us define

$$M^s = \sup_{x\geq 0} (\underline{u}(x) - \overline{u}^s(x)), \tag{4.43}$$

as in the proof of Proposition 4.3, there exist $y_{\lambda,s} \geq x_{\lambda,s} > 0$ for any $s \in [1, 1+\varepsilon]$ such that

$$\underline{u}(x) - \Phi^\lambda(x, y_{\lambda,s}) \leq \underline{u}(x_{\lambda,s}) - \Phi^\lambda(x_{\lambda,s}, y_{\lambda,s}), \tag{4.44}$$

where Φ^λ is defined in (4.5). Let m be a common Lipschitz constant for $\underline{u}$ and $\overline{u}^s$ with $s \in [1, 1+\varepsilon]$.

Assume first that the functions $\underline{u}$ and $\overline{u}^s$ are twice continuously differentiable at $x_{\lambda,s}$ and $y_{\lambda,s}$, respectively. Since $\Sigma_{\lambda,s}$ [as defined in (4.6)] reaches a local maximum at $(x_{\lambda,s}, y_{\lambda,s})$, we have that $\Phi_x^\lambda(x_{\lambda,s}, y_{\lambda,s}) = \underline{u}'(x_{\lambda,s})$, $-\Phi_y^\lambda(x_{\lambda,s}, y_{\lambda,s}) = (\overline{u}^s)'(y_{\lambda,s})$. Defining $A = \underline{u}''(x_{\lambda,s})$ and $B = (\overline{u}^s)''(y_{\lambda,s})$, we obtain

$$\left(\Phi_x^\lambda(x_{\lambda,s}, y_{\lambda,s}), A\right) \in D_2^+(\underline{u})(x_{\lambda,s}),$$

$$\left(-\Phi_y^\lambda(x_{\lambda,s}, y_{\lambda,s}), B\right) \in D_2^-(\overline{u}^s)(y_{\lambda,s}),$$

and

$$\begin{pmatrix} A & 0 \\ 0 & -B \end{pmatrix} \leq D^2\Phi^\lambda(x_{\lambda,s}, y_{\lambda,s}),$$

where $D^2\Phi^\lambda$ is the matrix of second derivatives of Φ^λ, the matrix inequality means that the difference is a negative semi-definite matrix, and D_2^+ and D_2^- are defined in Definition 3.6.

In the case that $\underline{u}$ and $\overline{u}^s$ are not both twice continuously differentiable at $x_{\lambda,s}$ and $y_{\lambda,s}$, respectively, we can resort to a more general theorem to get a similar result.

Using Theorem 3.2 of Crandall et al. [22], it can be proved that for any $\delta > 0$, there exist real numbers A_δ and B_δ such that

$$\begin{aligned} &\left(\Phi_x^\lambda(x_{\lambda,s}, y_{\lambda,s}), A_\delta\right) \in D_2^+(\underline{u})(x_{\lambda,s}), \\ &\left(-\Phi_y^\lambda(x_{\lambda,s}, y_{\lambda,s}), B_\delta\right) \in D_2^-(\overline{u}^s)(y_{\lambda,s}), \end{aligned} \tag{4.45}$$

and

$$\begin{pmatrix} A_\delta & 0 \\ 0 & -B_\delta \end{pmatrix} \leq D^2\Phi^\lambda(x_{\lambda,s}, y_{\lambda,s}) + \delta\left(D^2\Phi^\lambda(x_{\lambda,s}, y_{\lambda,s})\right)^2. \tag{4.46}$$

Let us call

$$\mathcal{L}_\gamma(\underline{u}, d, q)(x) = \tfrac{\sigma^2\gamma^2x^2}{2}q + (p + r\gamma x)\,d - \beta u(x) + \beta\mathcal{I}(u)(x).$$

So, we obtain from (3.7) and (4.45)

$$\sup\nolimits_{\gamma\in\Gamma} \mathcal{L}_\gamma(\underline{u}, \Phi_x^\lambda(x_{\lambda,s}, y_{\lambda,s}), A_\delta)(x_{\lambda,s}) \geq 0 \tag{4.47}$$

and

$$\sup\nolimits_{\gamma\in\Gamma} \mathcal{L}_\gamma(\overline{u}^s, -\Phi_y^\lambda(x_{\lambda,s}, y_{\lambda,s}), B_\delta)(y_{\lambda,s}) \leq 0. \tag{4.48}$$

From (4.46), we get

$$\begin{aligned} &A_\delta x_{\lambda,s}^2 - B_\delta y_{\lambda,s}^2 \\ &\leq \left(\left(\lambda + \tfrac{4m\lambda}{\left(\lambda\left(y_{\lambda,s}-x_{\lambda,s}\right)+1\right)^3}\right) + 2\delta\left(\lambda + \tfrac{4m\lambda}{\left(\lambda\left(y_{\lambda,s}-x_{\lambda,s}\right)+1\right)^3}\right)^2\right)(x_{\lambda,s} - y_{\lambda,s})^2. \end{aligned} \tag{4.49}$$

By (4.5), we have

$$\Phi_x^\lambda(x_{\lambda,s}, y_{\lambda,s}) + \Phi_y^\lambda(x_{\lambda,s}, y_{\lambda,s}) = 0 \tag{4.50}$$

and

$$\begin{aligned} &x_{\lambda,s}\Phi_x^\lambda(x_{\lambda,s}, y_{\lambda,s}) + y_{\lambda,s}\Phi_y^\lambda(x_{\lambda,s}, y_{\lambda,s}) \\ &= \lambda(x_{\lambda,s} - y_{\lambda,s})^2 + \tfrac{2m}{\left(\lambda\left(y_{\lambda,s}-x_{\lambda,s}\right)+1\right)^2}(x_{\lambda,s} - y_{\lambda,s}). \end{aligned} \tag{4.51}$$

But $\left(-\Phi_y^\lambda(x_{\lambda,s}, y_{\lambda,s}), B_\delta\right) \in D_2^-(\overline{u}^s)(y_{\lambda,s})$, so we obtain that $-\Phi_y^\lambda(x_{\lambda,s}, y_{\lambda,s}) \geq s > 1$. From (4.47) and (4.50) we conclude

$$\sup_{\gamma\in\Gamma} \mathcal{L}_\gamma(\underline{u}, \Phi_x^\lambda(x_{\lambda,s}, y_{\lambda,s}), A_\delta)(x_{\lambda,s}) \geq 0. \tag{4.52}$$

Therefore, taking $\gamma_{\lambda,s} = \arg\max_{\gamma\in\Gamma} \mathcal{L}_\gamma(\underline{u}, \Phi_x^\lambda(x_{\lambda,s}, y_{\lambda,s}), A_\delta)(x_{\lambda,s})$, we get from (4.48) and (4.52)

$$0 \le \mathcal{L}_{\gamma_{\lambda,s}}(\underline{u}, \Phi_x^\lambda(x_{\lambda,s}, y_{\lambda,s}), A_\delta)(x_{\lambda,s}) - \mathcal{L}_{\gamma_{\lambda,s}}(\overline{u}^s, -\Phi_y^\lambda(x_{\lambda,s}, y_{\lambda,s}), B_\delta)(y_{\lambda,s}).$$

So

$$\begin{array}{l}\beta\left(\underline{u}(x_{\lambda,s}) - \overline{u}^s(y_{\lambda,s})\right)\\ \le \frac{\sigma^2\gamma_{\lambda,s}^2}{2}(A_\delta x_{\lambda,s}^2 - B_\delta y_{\lambda,s}^2)\\ \quad + p(\Phi_x^\lambda(x_{\lambda,s}, y_{\lambda,s}) + \Phi_y^\lambda(x_{\lambda,s}, y_{\lambda,s}))\\ \quad + r\gamma_{\lambda,s}(\Phi_x^\lambda(x_{\lambda,s}, y_{\lambda,s})x_{\lambda,s} + \Phi_y^\lambda(x_{\lambda,s}, y_{\lambda,s})y_{\lambda,s})\\ \quad + \beta\left(\int_0^{x_{\lambda,s}} \underline{u}(x_{\lambda,s} - \alpha)dF(\alpha) - \int_0^{y_{\lambda,s}} \overline{u}^s(y_{\lambda,s} - \alpha)dF(\alpha)\right).\end{array} \tag{4.53}$$

Using the inequality

$$\Sigma_{\lambda,s}(x_{\lambda,s}, x_{\lambda,s}) + \Sigma_{\lambda,s}(y_{\lambda,s}, y_{\lambda,s}) \le 2\Sigma_{\lambda,s}(x_{\lambda,s}, y_{\lambda,s})$$

we have

$$\lambda\,(x_{\lambda,s} - y_{\lambda,s})^2 \le \underline{u}(x_{\lambda,s}) - \underline{u}(y_{\lambda,s}) + \overline{u}^s(x_{\lambda,s}) - \overline{u}^s(y_{\lambda,s}) + 4m(y_{\lambda,s} - x_{\lambda,s}).$$

Then, we get that

$$\lambda\,(x_{\lambda,s} - y_{\lambda,s})^2 \le 6m\,|x_{\lambda,s} - y_{\lambda,s}|. \tag{4.54}$$

We can find a sequence $\lambda_{n,s} \to \infty$ such that $(x_{\lambda_n,s}, y_{\lambda_n,s}) \to (\overline{x}_s, \overline{y}_s) \in A$. From (4.54), we get that $|x_{\lambda_n,s} - y_{\lambda_n,s}| \le 6m/\lambda_{n,s}$ and this gives $\overline{x}_s = \overline{y}_s$ and so $\lim_{n\to\infty} \lambda_{n,s}(x_{\lambda_n,s} - y_{\lambda_n,s})^2 = 0$. Taking $\delta_s = 1/\lambda_s$, and using that $y_{\lambda_n,s} \ge x_{\lambda_n,s}$ for all n, (4.49)–(4.51) and (4.53), we get

$$\beta\left(\underline{u}(\overline{x}_s) - \overline{u}^s(\overline{x}_s)\right) \le \beta \int_0^{C^s} \left(\underline{u}(\overline{x}_s - \alpha) - \overline{u}^s(\overline{x}_s - \alpha)\right) dF(\alpha), \tag{4.55}$$

where C^s is equal to either $\overline{x}_s$ or $\overline{x}_s^-$. From (4.8) and (4.55) we obtain

$$\begin{array}{rl} M^s & \le \lim_{n\to\infty} \Sigma_{\lambda,s}(x_{\lambda_n,s}, y_{\lambda_n,s})\\ & = \underline{u}(\overline{x}_s) - s\overline{u}(\overline{x}_s)\\ & \le \int_0^{C^s} \left(\underline{u}(\overline{x}_s - \alpha) - s\overline{u}(\overline{x}_s - \alpha)\right) dF(\alpha)\\ & \le \int_0^{C^s} M^s dF(\alpha)\\ & \le M^s\, F(C^s)\\ & \le M^s. \end{array} \tag{4.56}$$

If $F(C^s) < 1$ for some $s \in (1, 1+\varepsilon]$ there is a contradiction and so we have the result. If this is not the case, $F(\overline{x}_s) = 1$ for all $s \in (1, 1+\varepsilon]$ and F is absolutely continuous. Let us define

$$H = \left\{x \ge 0 : F'(x) > 0\right\}.$$

Then, H is a bounded set by x_s and it has positive Lebesgue measure. From (4.56),

$$\underline{u} - s\overline{u} = \underline{u}(\overline{x}_s) - s\overline{u}(\overline{x}_s) = M^s \text{ a.e. in } I_s := x_s - H.$$

Since the set $(1, 1+\varepsilon]$ is not countable, there exist $s_1 > s_2$ in $(1, 1+\varepsilon]$ such that $I = I_{s_1} \cap I_{s_2}$ has positive Lebesgue measure. So, $\underline{u} - s_1\overline{u} = M_{s_1}$ and $\underline{u} - s_2\overline{u} = M_{s_2}$ a.e. in I and we have that

$$\underline{u} = \frac{s_1 M_{s_2} - s_2 M_{s_1}}{s_1 - s_2} \text{ and } \overline{u} = \frac{M_{s_2} - M_{s_1}}{s_1 - s_2} \text{ a.e. in } I.$$

Both $\underline{u}$ and $\overline{u}$ are nondecreasing, and then there exists an interval $(a, b) \supset I$ such that

$$\underline{u}(x) = \frac{s_1 M_{s_2} - s_2 M_{s_1}}{s_1 - s_2} \text{ and } \overline{u}(x) = \frac{M_{s_2} - M_{s_1}}{s_1 - s_2} \text{ with } x \in (a, b).$$

Therefore, for all $x \in (a, b)$, we have

$$\begin{aligned} 0 \leq \sup_{\gamma \in \Gamma} \mathcal{L}_\gamma(\underline{u})(x) &= -\beta \underline{u}(x) + \beta \mathcal{I}(\underline{u})(x) \\ &\leq \beta(F(x) - 1)\underline{u}(x) \leq 0. \end{aligned}$$

If $F(x) < 1$ for some $x \in (a, b)$, we have a contradiction. If $F(x) = 1$ for all $x \in (a, b)$, so $F(a) = 1$ and

$$-\beta \underline{u}(a) + \beta \mathcal{I}(\underline{u})(a) = 0.$$

So, since $\underline{u}$ is nondecreasing,

$$-\beta \underline{u}(x) + \beta \mathcal{I}(\underline{u})(x) = 0$$

for all $x < a$. But there exists $x_2 \in (0, a)$ with $F(x_2) < 1$, and then

$$0 = -\beta \underline{u}(x_2) + \beta \mathcal{I}(\underline{u})(x_2) \leq \beta(F(x_2) - 1)\underline{u}(x_2) < 0$$

which is a contradiction. □

From the previous Proposition and Remark 3.5 we obtain the following result.

Theorem 4.5. *The function defined in (2.28) can be characterized as the unique viscosity solution of (2.30) with limit one at infinity.*

4.6 Dividends and Investments

As we did in Sect. 4.2, we use a comparison-principle approach to address the issue of uniqueness of viscosity solutions for the HJB equation (2.42) satisfying growth condition A.1 and a boundary condition at zero. We also prove a characterization result.

Proposition 4.6. *Let us assume that for all $x > 0$, $\underline{u}(x)$ is a subsolution of (2.42) and $\overline{u}(x)$ is a supersolution of (2.42) both nondecreasing and nonnegative functions satisfying the growth condition A.1. If $\underline{u}(0) = \overline{u}(0)$, then $\underline{u} \le \overline{u}$ in $\mathbf{R}_+$.*

Proof. The argument is the same as the one of Proposition 4.3. The only difference is that, since $\sup_{\gamma\in\Gamma} \tilde{\mathcal{L}}_\gamma$ is a second-order operator, it is necessary to show, as in the proof of Proposition 4.5, that there exist real numbers A_δ and B_δ such that $\left(\Phi_x^\lambda(x_{\lambda,s}, y_{\lambda,s}), A_\delta\right) \in D_2^+ (\underline{u}) (x_\lambda)$ and $\left(-\Phi_y^\lambda(x_{\lambda,s}, y_{\lambda,s}), B_\delta\right) \in D_2^- (\overline{u}^s) (y_{\lambda,s})$ for any $\delta > 0$. □

The next corollary is a direct consequence of the previous proposition.

Corollary 4.2. *There is at most one viscosity solution of (2.42) with boundary condition $u(0) = u_0$ among all the functions that satisfy growth condition A.1.*

In order to prove the characterization result, we have to prove two technical lemmas, the first one is a version of Lemma 4.1 for this case.

Lemma 4.5. *Let $\overline{u}$ be an absolutely continuous nonnegative supersolution of (2.42) in $(0, +\infty)$. Given any pair of real numbers $x_1 > x_0 > 0$, we can find a sequence of nonnegative functions $u_n : \mathbf{R} \to \mathbf{R}$ such that:*

(a) u_n is twice continuously differentiable.
(b) u_n converges uniformly to $\overline{u}$ in $[0, x_1]$.
(c) $u_n'(x) \ge 1$ in $[x_0, x_1]$.
(d) $\limsup_{n\to\infty} \left(\sup_{\gamma\in\Gamma} \tilde{\mathcal{L}}_\gamma(u_n)(x) \right) \le 0$ in $[x_0, x_1]$.

Proof. The proofs of (a) to (c) are similar to the ones in Lemma 4.1 but taking the left-side convolution $u_n = \overline{u} * \phi_n$ where $\phi_n(x) = n\phi(n(x - 1/n))$ and ϕ is an even and twice continuously differentiable function with support included in $(-1, 1)$, with integral one, satisfying $\phi' \ge 0$ in $(-1, 0)$ and $\phi' \le 0$ in $(0, 1)$.

Let us prove (d). By Proposition 3.4, $\overline{u}$ is semiconcave and so $\overline{u}''$ exists a.e. and the possible jumps of $\overline{u}$ are downward. So, the left-sided convolution u_n satisfies $u_n''(x) \le \left(\overline{u}'' * \overline{\phi}_n\right)(x)$. The result (d) follows because $\tilde{\mathcal{L}}_\gamma(\overline{u}) \le 0$ for any $\gamma \in [0, 1]$ a.e. in $[x_0, x_1]$, and it can be shown that

$$\limsup_{n\to\infty} \left(\tilde{\mathcal{L}}_\gamma(u_n)(x) - \left(\tilde{\mathcal{L}}_\gamma(\overline{u}) * \overline{\phi}_n\right)(x)\right) \le 0$$

for all $x \in [x_0, x_1]$. □

In the second technical lemma, we show that the optimal value function (2.33) can be written as a limit of value functions of strategies whose surpluses are confined in compact subsets of $(0,\infty)$. See Lemma A.1 in [11] for the details of the proof.

Lemma 4.6. *Given $x \geq 0$ and $x_1 > x$, let us define $\Pi_x^{x_1}$ as the set of $(\overline{L},\overline{\gamma}) \in \Pi_x^{L,\gamma}$ such that $X_t^{\overline{L},\overline{\gamma}} \leq x_1$ for all $t \geq 0$ and $\mathcal{V}^{x_1}(x) = \sup\{V_{\overline{L},\overline{\gamma}}(x)$ with $(\overline{L},\overline{\gamma}) \in \Pi_x^{x_1}\}$, then*

$$\lim_{x_1\to\infty} \mathcal{V}^{x_1}(x) = V(x).$$

The proof that the optimal value function (2.33) is the smallest supersolution of (2.42) is similar to the one of Proposition 4.4, but in this case we should also consider a martingale that involves the Brownian motion; we use the previous lemma in order to confine the controlled surplus process to a compact set.

Proposition 4.7. *The optimal value function defined in (2.33) is the smallest viscosity supersolution of (2.42).*

Proof. Taking $x > 0$, by Lemma 4.6, it is enough to proof that for any pair (x_0, x_1) such that $0 < x_0 \leq x \leq x_1$, we have that

$$\sup_{(\overline{L},\overline{\gamma})\in\Pi_x^{[x_0,x_1]}} V_{\overline{L},\overline{\gamma}}(x) \leq \overline{u}(x),$$

where $\Pi_x^{[x_0,x_1]}$ is the set of admissible strategies $(\overline{L},\overline{\gamma}) \in \Pi_x^{x_1}$ such that no dividends are paid after

$$\tau_{x_0} := \min\{t : X_t^{\overline{L},\overline{\gamma}} < x_0\}.$$

Take $(\overline{L},\overline{\gamma}) \in \Pi_x^{[x_0,x_1]}$ and let us define X_t as the corresponding controlled surplus process and τ as the ruin time. Consider the twice continuously differentiable functions u_n defined in Lemma 4.5 extended as $u_n(x) = 0$ for $x < 0$; by Proposition 2.13 and since $u_n' \geq 1$, we can write

$$u_n(X_{t\wedge\tau_{x_0}})e^{-c(t\wedge\tau_{x_0})} - u_n(x) \leq \int_0^{t\wedge\tau_{x_0}} \tilde{\mathcal{L}}_{\gamma_s}(u_n)(X_{s^-})e^{-cs}ds - \int_0^{t\wedge\tau_{x_0}} e^{-cs}dL_s + \tilde{M}_{t\wedge\tau_{x_0}}, \tag{4.57}$$

where $\tilde{M}_{t\wedge\tau_{x_0}}$ is a martingale with zero expectation because $X_{s^-} \in [x_0, x_1]$ for $s \leq \tau_{x_0}$. We have

$$E_x\left(u_n(X_{t\wedge\tau_{x_0}})e^{-c(t\wedge\tau_{x_0})}\right) \geq 0 \tag{4.58}$$

and from Lemma 4.5(d), we also have

$$\limsup_{n\to\infty} E_x\left(\int_0^{t\wedge\tau_{x_0}} \tilde{\mathcal{L}}_{\gamma_s}(u_n)(X_{s^-})e^{-cs}ds\right) \leq 0 \tag{4.59}$$

for all t. So, from Lemma 4.5(b), (4.57)–(4.59), we obtain

$$\overline{u}(x) = \lim_{n\to\infty} u_n(x) \geq E_x(\int_0^{t\wedge\tau_{x_0}} e^{-cs} dL_s) = E_x(\int_0^{t\wedge\tau} e^{-cs} dL_s).$$

Since

$$\lim_{t\to\infty} E_x\left(\int_0^{t\wedge\tau} e^{-cs} dL_s\right) = V_{\overline{L},\overline{\gamma}}(x) \tag{4.60}$$

and the function V defined in (2.33) is a viscosity solution of (2.42), we have the result. □

As we discussed in Remark 2.7, we do not know yet whether the optimal value function (2.33) satisfies the growth condition A.1; but, using the fact that the linear functions $u(x) = x + k$ with $k > p/c$ are viscosity supersolutions of (2.42), we conclude, by the previous proposition, that the optimal value function satisfies this growth condition.

From the definition of (2.33) and from Proposition 4.7 we get the following verification result.

Theorem 4.6. *Consider a family of admissible strategies* $\left(\left(\overline{L}^x, \gamma^x\right)\right)_{x\geq 0}$ *such that* $\left(\overline{L}^x, \gamma^x\right) \in \Pi_x^{L,\gamma}$ *for any* $x \geq 0$*. If the function* $V_{\left(\overline{L}^x,\gamma^x\right)}(x)$ *is a viscosity supersolution of (2.42), then* $V_{\left(\overline{L}^x,\gamma^x\right)}(x)$ *is the optimal value function (2.33).*

Chapter 5
Optimal Strategies

The aim of the present chapter is to show the existence of optimal stationary strategies in the classical risk models. We start with the problems of dividend payments and consider first the simplest problem without reinsurance or investment control.

5.1 Dividend Band Strategies

We define a particular stationary strategy where the dividends are paid in three different ways: A dividend band strategy is characterized by three sets $\mathcal{A}$, $\mathcal{B}$, and $\mathcal{C}$ which partitioned the state space of the surplus process. If the current surplus is in $\mathcal{A}$ (continuous-payment set), all the incoming premium is paid as dividends; if the current surplus is in $\mathcal{B}$ (impulse-payment set), a positive amount of money is paid as dividends in order to bring the surplus process back to $\mathcal{A}$; and finally if the surplus is in $\mathcal{C}$ (no action set), no dividends are paid. The band strategies are *stationary* in the sense that they only depend on the current surplus. We will show in the next section that the optimal dividend strategies of problem (1.10) are indeed band strategies.

Definition 5.1. We say that $\mathcal{P} = (\mathcal{A}, \mathcal{B}, \mathcal{C})$ is a *band partition* if $\mathcal{A}$, $\mathcal{B}$, and $\mathcal{C}$ are disjoint sets with $\mathbf{R}_+ = \mathcal{A} \cup \mathcal{B} \cup \mathcal{C}$, $\mathcal{A}$ is closed, bounded, and nonempty; $\mathcal{B}$ is left open; $\mathcal{C}$ is right open; the lower limit of any connected component of $\mathcal{B}$ belongs to $\mathcal{A}$; and there exists $b \geq 0$ such that $(b, \infty) \subset \mathcal{B}$.

Remark 5.1. This definition implies that both $\mathcal{A}$ and $\mathcal{B}$ are nonempty. The point $y = \inf \mathcal{B}$ cannot belong to $\mathcal{B}$ because $\mathcal{B}$ is left open and cannot belong to $\mathcal{C}$ because $\mathcal{C}$ is right open, so $y \in \mathcal{A}$. However, $\mathcal{C}$ could be empty.

Let us define the band strategy associated to a band partition.

Definition 5.2. Given a band partition $\mathcal{P} = (\mathcal{A}, \mathcal{B}, \mathcal{C})$ and an initial surplus $x \geq 0$, we define recursively the admissible strategy $\overline{L}^x = (L_t^x)_{t \geq 0} \in \Pi_x^L$ as follows:

P. Azcue and N. Muler, *Stochastic Optimization in Insurance: A Dynamic Programming Approach*, SpringerBriefs in Quantitative Finance, DOI 10.1007/978-1-4939-0995-7_5,

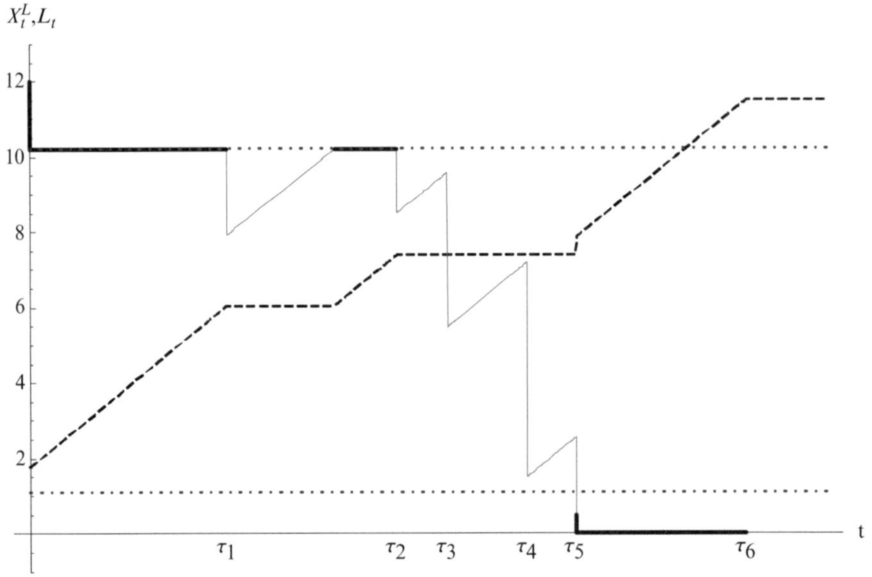

Fig. 5.1 Dividend strategy and controlled surplus under a band strategy

- In the case that $x \in \mathcal{A}$, then $L_t^x = pt$ up to the first-claim arrival τ_1. Afterwards, follow the strategy corresponding to initial surplus $x - U_1$ where U_1 is the size of the first claim.
- In the case that $x \in \mathcal{B}$, there exists an open interval $(x_0, x) \subset \mathcal{B}$ with $x_0 \in \mathcal{A}$; then $L_0^x = x - x_0$. Afterwards, follow the strategy corresponding to initial surplus $x_0 \in \mathcal{A}$.
- In the case that $x \in \mathcal{C}$, there exists an open interval $(x, x_1) \subset \mathcal{C}$ with $x_1 \in \mathcal{A}$; take $L_t^x = 0$ up to $\tau_{\mathcal{C}}$ the exit time of $\mathcal{C}$. Afterwards, follow the strategy corresponding to initial surplus $X_{\tau_{\mathcal{C}}}^{\overline{L}^x}$.

The family $\pi(\mathcal{P}) = \{\overline{L}^x \in \Pi_x^L \text{ with } x \geq 0\}$ is called the *band strategy* associated to the band partition $\mathcal{P}$.

Let us consider, for example, the band partition $\mathcal{A} = \{0, 10.22\}$, $\mathcal{B} = (0, 1.083] \cup (10.22, \infty)$, and $\mathcal{C} = (1.083, 10.22)$. In Fig. 5.1, we show the dividend strategy $\overline{L} = (L_t)_{t \geq 0}$ in dashed line and the controlled surplus $X_t^{\overline{L}}$ in solid line for initial surplus $x = 12$ and a particular occurrence of claims (the dotted lines indicate the boundaries of the set $\mathcal{C}$); the bold part of the graph of $X_t^{\overline{L}}$ indicates that dividends are paid. This band partition corresponds to the optimal one of the first example which we present in Sect. 6.2.1.

Remark 5.2. The simplest band strategy is the so-called *barrier* strategy: given a certain barrier level $a \geq 0$, if the current surplus x exceeds a, a positive amount of money $x - a$ is paid as dividends, no dividends is paid when the surplus is less than a, and if the surplus is a, all the incoming premium is paid as dividends. More precisely, $\mathcal{A} = \{a\}$, $\mathcal{B} = (a, \infty)$, and $\mathcal{C} = [0, a)$.

We define the value function $V_{\mathcal{P}} : \mathbf{R}_+ \to \mathbf{R}$ of the band strategy $\pi(\mathcal{P}) = \{\overline{L}^x \in \Pi_x^L \text{ with } x \geq 0\}$ associated to the band partition $\mathcal{P}$ as $V_{\mathcal{P}}(x) = V_{\overline{L}^x}(x)$. The following proposition show that the function $V_{\mathcal{P}}$ is the fixed point of a suitable operator.

Proposition 5.1. *Let $\pi(\mathcal{P}) = \{\overline{L}^x \in \Pi_x^L$ with $x \geq 0\}$ be the band strategy associated to the band partition $\mathcal{P}$. Let us define the operator T on the set of nonnegative, Borel-measurable, and locally bounded functions as*

$$\mathcal{T}(W)(x) = E_x\left(\int_0^{\tau_1} e^{-cs} dL_s^x + e^{-c\tau_1} W(X_{\tau_1}^{\overline{L}^x})\right),$$

where τ_1 and U_1 are the time and size of the first claim, respectively. Then, the value function $V_{\mathcal{P}}$ is the unique fixed point of the operator $\mathcal{T}$.

Proof. Given an initial surplus $x \geq 0$, we have that

$$X_t^{\overline{L}^x} \leq m(x) := \inf\{y \geq x \text{ such that } y \in \mathcal{A} \cup \mathcal{B}\};$$

$m(x)$ is finite because $m(x) = x$ if $x \in \mathcal{A} \cup \mathcal{B}$, and if $x \in \mathcal{C}$, $m(x)$ is the upper bound of the connected component of $\mathcal{C}$ where x lies. Given any $M \in \mathcal{A} \cup \mathcal{B}$, consider the complete metric space

$$B_M = \{W : [0, M] \to \mathbf{R}_+ \text{ Borel-measurable and bounded}\}$$

with the metric $d(W_1, W_2) = \sup_{x \in [0,M]} |W_1(x) - W_2(x)|$. Note that for any $x \leq M$, we have $m(x) \leq M$ and so $\mathcal{T}(W)$ is well defined and bounded in $[0, M]$. On the other hand,

$$|\mathcal{T}(W_1)(x) - \mathcal{T}(W_2)(x)| \leq E_x(e^{-c\tau_1})\, d(W_1, W_2) = \tfrac{\beta}{c+\beta} d(W_1, W_2),$$

so $\mathcal{T} : B_M \to B_M$ is a contraction operator with modulus $\beta/(c + \beta) < 1$ and therefore it has a unique fixed point. Since any connected component of $\mathcal{C}$ is bounded we have that if $\sup \mathcal{A} \cup \mathcal{B} = \infty$, then there exists a unique locally bounded function $W^* : \mathbf{R}_+ \to \mathbf{R}_+$ with $T(W^*) = W^*$.

Since $T(V_{\mathcal{P}}) = V_{\mathcal{P}}$, we obtain that $V_{\mathcal{P}}$ is the unique fixed point. □

Let us consider the operator Λ defined as

$$\Lambda(W)(x) = p - (c + \beta)W(x) + \beta\mathcal{I}(W)(x) \tag{5.1}$$

and the operator $\tilde{\mathcal{L}}_0$ defined in (1.22). The following proposition is a verification result: a function W which is obtained gluing in a certain way linear functions of slope one, almost-everywhere solutions of the equations $\tilde{\mathcal{L}}_0 = 0$ and solutions of $\Lambda = 0$, turns to be the value function of the band strategy.

Proposition 5.2. *Consider a band partition $\mathcal{P} = (\mathcal{A}, \mathcal{B}, \mathcal{C})$; any Borel-measurable function W which is left continuous at the upper limit of the connected component of $\mathcal{C}$, it is right continuous at the lower limits of the connected components of $\mathcal{B}$, it has derivative equal to 1 on $\mathcal{B}$, it is an almost-everywhere solution of $\tilde{\mathcal{L}}_0 = 0$ in the connected components of $\mathcal{C}$, and it is a solution of $\Lambda = 0$ in $\mathcal{A}$, should be the value function $V_{\mathcal{P}}$.*

Proof. By Lemma 5.1, it is enough to show that $\mathcal{T}(W) = W$. If $x \in \mathcal{A}$, then $\Lambda(W)(x) = 0$ and so

$$\begin{aligned}\mathcal{T}(W)(x) &= E_x\left(\int_0^{\tau_1} e^{-cs} p ds + e^{-c\tau_1} W(x)\right)\\ &= \tfrac{1}{c+\beta}\left(p - \beta\mathcal{I}(W)(x)\right)\\ &= W(x).\end{aligned}$$

If $x \in \mathcal{B}$, consider $x_0 = \max\{y < x \text{ and } y \notin \mathcal{B}\}$. We have that $x_0 \in \mathcal{A}$ and so

$$\begin{aligned}\mathcal{T}(W)(x) &= E_x\left(x - x_0 + \int_0^{\tau_1} e^{-cs} p ds + e^{-c\tau_1} W(x_0)\right)\\ &= x - x_0 + W(x_0)\\ &= W(x).\end{aligned}$$

Finally, if $x \in \mathcal{C}$, consider $x_1 = \min\{y > x \text{ and } y \notin \mathcal{C}\}$, $t_1 = (x_1 - x)/p$, and $x(t) = x + pt$. The set

$$\mathcal{E} = \{t \in [0, t_1] : W \text{ is differentiable at } x + pt\}$$

has full measure. We have that $x(t) \in \mathcal{C}$ for $t \in [0, t_1)$ and $x(t_1) = x_1 \in \mathcal{A}$. So, $\mathcal{T}(W)(x_1) = W(x_1)$ and

$$\begin{aligned}&\mathcal{T}(W)(x)\\ &= E_x\left(I_{\tau_1 > t_1} e^{-ct_1} W(x_1)\right) + E_x\left(I_{\tau_1 \le t_1} e^{-c\tau_1} W(x(\tau_1) - U_1)\right)\\ &= e^{-(c+\beta)t_1} W(x_1) + \int_0^{t_1}\left(\int_0^{x(t)} W(x(t) - \alpha) dF(\alpha)\right)\beta e^{-(c+\beta)t} dt\\ &= e^{-(c+\beta)t_1} W(x_1) + \int_0^{t_1} I_{t\in\mathcal{E}}\left(pW'(x(t)) - (c+\beta) W(x(t))\right)e^{-(c+\beta)t} dt\\ &= e^{-(c+\beta)t_1} W(x_1) + \int_0^{t_1} I_{t\in\mathcal{E}}\left(-W(x(t))e^{-(c+\beta)t}\right)' dt\\ &= W(x).\end{aligned}$$

□

5.2 Optimal Dividend Strategies

In this section we prove that the optimal value function V defined in (1.10) is the value function of a band strategy $\pi(\mathcal{P}^*)$ associated to some band partition $\mathcal{P}^*$. We will show that the optimal band partition $\mathcal{P}^*$ can be obtained from V as follows.

Definition 5.3. Let us define $\mathcal{P}^* = (\mathcal{A}^*, \mathcal{B}^*, \mathcal{C}^*)$ as

- $\mathcal{A}^* = \{x \in \mathbf{R}_+ \text{ such that } \Lambda(V)(x) = 0\}$.
- $\mathcal{B}^* = \{x \in (0, \infty) \text{ such that } V'(x) = 1 \text{ and } \Lambda(V)(x) < 0\}$.
- $\mathcal{C}^* = (\mathcal{A}^* \cup \mathcal{B}^*)^c$.

Our first goal is to prove that $\mathcal{P}^*$ satisfies the properties of a band partition. In order to do that, we state a proposition that gives conditions under which the optimal value function is the supremum of the value functions corresponding to admissible strategies with surplus not exceeding some threshold $\hat{x}$. The argument relies on the idea that at a point $\hat{x}$ such that either $\Lambda(V)(\hat{x}) = 0$ or $V'(\hat{x}) = 1$ the best strategy (if it exists) would be either to pay the incoming premium as dividends or to make a lump payment of dividends. In each of these cases the controlled surplus process would not surpass the threshold $\hat{x}$. The proof is very technical.

Proposition 5.3. *Let us define for any $\hat{x} \geq x$, the set*

$$\Pi^L_{x,\hat{x}} = \{\overline{L} \in \Pi^L_x \text{ such that } X^{\overline{L}}_t \leq \hat{x} \text{ for all } t \geq 0\}.$$

Take $\hat{x} \geq 0$ satisfying either $\Lambda(V)(\hat{x}) = 0$ or $V'(\hat{x}) = 1$, and then for any $x \leq \hat{x}$,

$$V(x) = \sup\nolimits_{\overline{L} \in \Pi^L_{x,\hat{x}}} V_{\overline{L}}(x).$$

Proof. Let us consider the case where $\Lambda(V)(\hat{x}) = 0$; the proof for the case $V'(\hat{x}) = 1$ is similar and can be found in Lemma A.5 in [9].

We construct inductively a family of dividend strategies in $\Pi^L_{x,\hat{x}}$ pasting suitable dividend strategies. Since some measurability issues may occur with the pasting, we construct this strategies with detail. As a first step, we construct a family of admissible strategies $(\overline{L}^x \in \Pi^L_x)_{x \geq 0}$ such that $V(x) - V_{\overline{L}^x}(x) \leq \varepsilon/4$ for all $x \in [0, \hat{x})$. Consider m large enough such that the points $x_i = i\hat{x}/m$ with $i = 0, \ldots, m$ satisfy $V(x_i) - V(x_{i-1}) < \varepsilon/8$, and take admissible strategies $\overline{L}^{x_i} \in \Pi^L_{x_i}$ such that $V(x_i) - V_{\overline{L}^{x_i}}(x_i) < \varepsilon/8$. We define the admissible strategy $\overline{L}^x \in \Pi^L_x$ for any $x \in [0, \hat{x}]$ as follows: if $x \in [x_i, x_{i+1})$, pay immediately $x - x_i$ as dividends and then follow the strategy $\overline{L}^{x_i} \in \Pi^L_{x^i}$ up to the ruin time, so we have that $\overline{L}^x_t = x - x_i + \overline{L}^{x_i}_t$. For any $x \in [x_i, x_{i+1})$, we get

$$\begin{aligned} V(x) - V_{\overline{L}^x}(x) &= V(x) - \left((x - x_i) + V_{\overline{L}^{x_i}}(x_i)\right) \\ &\leq V(x_{i+1}) - V(x_i) + \left(V(x_i) - V_{\overline{L}^{x_i}}(x_i)\right) \\ &< \varepsilon/4. \end{aligned}$$

We have constructed these admissible strategies based on finite points in order to avoid measurability problems.

Let us define Π^n_x as the set of all the admissible strategies with initial surplus $x \leq \hat{x}$ whose surplus process does not exceed $\hat{x}$ before the arrival of the nth claim. Note that $\Pi^0 = \Pi^L_x$. As a first step we show that

$$V(x) = \sup\nolimits_{\overline{L} \in \Pi^n_x} V_{\overline{L}}(x). \tag{5.2}$$

The proof is by induction on n: By (1.10), the expression (5.2) holds for $n = 0$. Take $n \geq 1$ and $\varepsilon > 0$; we can find for each x_i with $i = 1, \ldots, m$, an admissible strategy $\overline{L}_{x_i}^{n-1} \in \Pi_{x_i}^{n-1}$ such that $V(x_i) - V_{\overline{L}_{x_i}^{n-1}}(x_i) < \frac{\varepsilon}{4}$; we will construct for any $x \leq \hat{x}$ an admissible strategy $\overline{L}_x^n = \left(L_t^n\right)_{t \geq 0} \in \Pi_x^n$ such that

$$0 \leq V_{\overline{L}_x^{n-1}}(x) - V_{\overline{L}_x^n}(x) < \frac{\varepsilon}{2}.$$

We now define the strategy $\overline{L}_x^n$: Starting with $x \leq \hat{x}$, follow the strategy $\overline{L}_x^{n-1}$ while $X_t^{\overline{L}_x^{n-1}} < \hat{x}$; if the surplus $X_t^{\overline{L}_x^{n-1}}$ reaches $\hat{x}$, pay out the incoming premium p as dividends up to next claim. If U is the size of this claim, follow the strategy $\overline{L}_{\hat{x}-U}^{n-1}$ (with initial surplus $\hat{x} - U$). Note that the strategy $\overline{L}_x^n = (L_t^{n,x})_{t \geq 0}$ is measurable (and then admissible).

In order to simplify the notation, hereafter we omit the value of the initial surplus in the notation of the dividend strategies. We show first that

$$V_{\overline{L}^n}(\hat{x}) \geq V_{\overline{L}^{n-1}}(\hat{x}) - \frac{\varepsilon}{2}. \tag{5.3}$$

If the initial surplus is $\hat{x}$, then $X_t^{\overline{L}^n} = \hat{x}$ and $L_t^n = pt$ for $t < \tau_1$ and $X_{\tau_1}^{\overline{L}^n} = \hat{x} - U_1$, where τ_1 and U_1 are the arrival time of and the size of the first claim, respectively. Since the probability of no claims in $[0, t]$ is $e^{-\beta t}$ and $\Lambda(V)(\hat{x}) = 0$, we have

$$\begin{aligned} V_{\overline{L}^n}(\hat{x}) &= E_{\hat{x}}(\textstyle\int_0^{\tau_1} e^{-cs} p ds) + E_{\hat{x}}(e^{-c\tau_1} V_{\overline{L}^{n-1}}(\hat{x} - U_1)) \\ &= \frac{1}{(c+\beta)} \left(p + \beta \textstyle\int_0^{\hat{x}} V_{\overline{L}^{n-1}}(\hat{x} - \alpha) dF(\alpha)\right) \\ &\geq \frac{1}{(c+\beta)} \left(p + \beta \textstyle\int_0^{\hat{x}} \left(V(\hat{x} - \alpha) - \frac{\varepsilon}{2}\right) dF(\alpha)\right) \\ &> V(\hat{x}) - \frac{\varepsilon}{2}. \end{aligned} \tag{5.4}$$

Now, given $x < \hat{x}$, let $\mathcal{M}$ be the set of all the paths with initial surplus x that reach $\hat{x}$ in finite time and let $\hat{\tau}$ be the first time that a path in $\mathcal{M}$ reaches $\hat{x}$. We can write

$$V_{\overline{L}^n}(x) = E_x(I_{\mathcal{M}} \textstyle\int_0^{\hat{\tau}} e^{-cs} dL_s^n) + E_x(I_{\mathcal{M}} e^{-c\hat{\tau}}) V_{\overline{L}^n}(\hat{x}) + E_x(I_{\mathcal{M}^c} \textstyle\int_0^{\tau^{\overline{L}^n}} e^{-cs} dL_s^n).$$

But since $X_t^{\overline{L}^n} = X_t^{\overline{L}^{n-1}}$ and $L_t^n = L_t^{n-1}$ in $\mathcal{M}^c$ for all t and also in $\mathcal{M}$ when $t < \hat{\tau}$, using (5.3), we obtain

$$V_{\overline{L}^{n-1}}(x) - V_{\overline{L}^n}(x) = E_x(I_{\mathcal{M}} e^{-c\hat{\tau}})(V_{\overline{L}^{n-1}}(\hat{x}) - V_{\overline{L}^n}(\hat{x})) \leq \frac{\varepsilon}{2} \tag{5.5}$$

for any $x < \hat{x}$. So, we get (5.2).

Finally, we prove that given any $\varepsilon > 0$ and any $x \in [0, \hat{x}]$, we can find an admissible strategy $\tilde{L} \in \Pi^L_{x,\hat{x}}$ such that

$$V(x) - V_{\tilde{L}}(x) < \varepsilon \text{ for all } x \in [0, \hat{x}].$$

Choose $t_1 > 0$ satisfying

$$e^{-ct_1} < \frac{\varepsilon}{4V(\hat{x})} \tag{5.6}$$

and $n \geq 1$ such that

$$P(N_{t_1} \geq n) = \sum_{k \geq n} \frac{e^{-\beta t_1} (\beta t_1)^k}{k!} < \frac{\varepsilon}{4V(\hat{x})}. \tag{5.7}$$

By (5.2), there exists $\overline{L}^n \in \Pi_n$ with

$$V_{\overline{L}^n}(x) \geq V(x) - \frac{\varepsilon}{2} \text{ for all } x \in [0, \hat{x}]. \tag{5.8}$$

Let us define

$$\tau^* = \inf\left\{t \geq 0 \colon X_t^{\overline{L}^n} > \hat{x}\right\}$$

and $\mathcal{N}$ the set of all the paths with initial surplus $x \in [0, \hat{x}]$ and τ^* finite. We define the strategy $\tilde{L}$ as follows: $\tilde{L} = \overline{L}^n$ for all $t < \tau^*$ and if $t = \tau^*$, pay out $\hat{x}$ immediately and then pay the incoming premium rate as dividends until the arrival of the next claim. Since $\overline{L}^n$ is admissible, it is easy to see that $\tilde{L}$ is also admissible. So, $\tilde{L} \in \Pi^L_{x,\hat{x}}$, and we have that

$$\begin{aligned} V_{\overline{L}^n}(x) &= E_x(I_{\mathcal{N}^c} \textstyle\int_0^{\tau^{\overline{L}^n}} e^{-cs} dL_s^n + I_{\mathcal{N}} \int_0^{\tau^*} e^{-cs} dL_s^n + I_{\mathcal{N}} e^{-c\tau^*} V_{\overline{L}^n}(\hat{x})) \\ &\leq E_x(I_{\mathcal{N}^c} \textstyle\int_0^{\tau^{\overline{L}^n}} e^{-cs} dL_s^n + I_{\mathcal{N}} \int_0^{\tau^*} e^{-cs} dL_s^n + I_{\mathcal{N}} e^{-c\tau^*} V(\hat{x})) \end{aligned}$$

And, since $\tilde{L} = \overline{L}^n$ in $\mathcal{N}^c$ for all t and also in $\mathcal{N}$ when $t < \tau^*$, we get

$$V_{\tilde{L}}(x) = E_x(I_{\mathcal{N}^c} \textstyle\int_0^{\tau^{\overline{L}^n}} e^{-cs} dL_s^n + I_{\mathcal{N}} \int_0^{\tau^*} e^{-cs} dL_s^n + I_{\mathcal{N}} e^{-c\tau^*} (\hat{x} + \frac{p}{c+\beta})).$$

Then, we conclude

$$\left| V_{\overline{L}^n}(x) - V_{\tilde{L}}(x) \right| \leq E_x(I_{\mathcal{N}} e^{-c\tau^*}) V(\hat{x}). \tag{5.9}$$

We have that $X^t_{\overline{L}^n}$ do not exceed $\hat{x}$ before the arrival of the nth claim, so $\{\tau^* < t_1\} \subset \{N_{t_1} \geq n\}$. Therefore, $\mathcal{N} = \{\tau^* < \infty\} \subset \{\tau^* \geq t_1\} \cup \{N_{t_1} \geq n\}$, and by (5.6) and (5.7) we obtain

$$
\begin{aligned}
E_x(I_{\mathcal{N}} e^{-c\tau^*}) &\le E_x(I_{\{\tau^* \ge t_1\}} e^{-c\tau^*}) + E_x(I_{\{N_{t_1} \ge n\}} e^{-c\tau^*}) \\
&\le e^{-ct_1} + P\left(\{N_{t_1} \ge n\}\right) \\
&< \frac{\varepsilon}{2V(\hat{x})}.
\end{aligned} \tag{5.10}
$$

From (5.9) and (5.10), we get

$$
\left| V_{\overline{L}^n}(x) - V_{\overline{L}}(x) \right| < \frac{\varepsilon}{2}. \tag{5.11}
$$

The result follows from (5.8) and (5.11). □

We also need the following three auxiliary lemmas.

Lemma 5.1. *The function* $\Lambda(V)$ *is right continuous with possible upward jumps and satisfies* $\Lambda(V) \le 0$.

Proof. Since V is locally Lipschitz and it is a viscosity solution of the HJB equation (1.21), then it satisfies this equation a.e.; so

$$
\Lambda(V) \le \tilde{\mathcal{L}}_0(V) \le 0 \text{ a.e.}
$$

Since F is right continuous with possible upward jumps, so is $\Lambda(V)$, then we have that $\Lambda(V) \le 0$ for all $x \ge 0$. □

Lemma 5.2. *Assume that there exists* $\hat{x} \ge 0$ *such that either* $\Lambda(V)(\hat{x}) = 0$ *or* $V'(\hat{x}) = 1$. *If a value function* W *of an admissible strategy (or a limit of value functions of admissible strategies) in* $\Pi^L_{x,\hat{x}}$ *is locally Lipschitz, nondecreasing, and a viscosity supersolution of (1.21) in* $[0, \hat{x}]$, *then* $W = V$ *in* $[0, \hat{x}]$.

Proof. With the same proof of Proposition 4.4, given any $\overline{L} \in \Pi^L_{x,\hat{x}}$ and any supersolution $\overline{u}$ of the HJB equation (1.21) in $[0, \hat{x}]$, we obtain that $V_{\overline{L}} \le \overline{u}$ in $[0, \hat{x}]$. So, by Lemma 5.3, we get the result. □

Lemma 5.3. *Consider a point* $y \ge 0$ *where* V *is differentiable; let us define the function* $W_y(x) = V(x)I_{\{x \le y\}} + (V(y) - y + x)I_{\{x > y\}}$. *Then,*

(a) $W_y \le V$.
(b) If W_y *is a viscosity supersolution of (1.21) in* (y, ∞), *then* $W_y = V$ *in* $\mathbf{R}_+$.
(c) Assume that either $\Lambda(V)(\hat{x}) = 0$ *or* $V'(\hat{x}) = 1$ *for some* $\hat{x} > 0$ *and there exists* $y < \hat{x}$ *such that* W_y *is a viscosity supersolution of (1.21) in* $(y, \hat{x}]$ *then* $W_y = V$ *in* $[0, \hat{x}]$.

Proof. (a) follows because $W_y(0) = V(0)$ and $W_y' \le V'$ almost everywhere.

In order to prove (b), let us show that $W_y \ge V$. By Proposition 4.4, it is enough to prove that W_y is supersolution of (1.21) in $[0, y]$. By definition, W_y is a viscosity solution of (1.21) in $[0, y)$. Let us see that W_y is a viscosity supersolution of (1.21) at y: it follows from Definition 3.2 that $D^-(W_y)(y)$ is not empty only in the case that $V'(y) = W_y'(y) = 1$ and since V is a viscosity solution at y, we have the result.

The proof of (c) is analogous to the proof of (a) but follows from Proposition 5.2. □

Proposition 5.4. *$\mathcal{P}^*$ is a band partition.*

Proof. From Lemma 5.1, we get that $\mathcal{A}^*$ is closed.

Let us prove that $\mathcal{B}^*$ is left open. Take $x \in \mathcal{B}^*$, and then by Lemma 5.1, there exists $\delta > 0$ such that $\Lambda(V)(y) < 0$ in $y \in [x-\delta, x]$ and V is differentiable at $x-\delta$. Consider the function $W_{x-\delta}$ introduced in Lemma 5.3; we have that $W'_{x-\delta}(y) = 1$ and $\tilde{\mathcal{L}}_0(W_{x-\delta})(y) = \Lambda(W_{x-\delta})(y) < 0$ for $y \in (x-\delta, x]$, and so $W_{x-\delta}$ is a supersolution of (1.21) in $(x-\delta, x]$. By Lemma 5.3(c), we get that $W_{x-\delta} = V$ in $(x-\delta, x]$ and then $(x-\delta, x] \subset \mathcal{B}^*$.

Let us show that the lower limit of any connected component of $\mathcal{B}^*$ lies in $\mathcal{A}^*$. Suppose first that $(0, x_1) \subset \mathcal{B}^*$; we are going to prove that $0 \in \mathcal{A}^*$. We are going to show that $V(0) = \max_{\overline{L} \in \Pi_0^L} V_{\overline{L}}(0) = p/(c+\beta)$; this implies that $\Lambda(V)(0) = 0$ and so $0 \in \mathcal{A}^*$. Consider the admissible strategy $\overline{L}^0 = \left(L_t^0\right)_{t \geq 0} \in \Pi_0^L$ with $L_t^0 = pt$ up to the first claim τ_1 (where the ruin occurs), then $V_{\overline{L}^0}(0) = p/(c+\beta)$. Take $x \in (0, x_1)$; since $V'(x) = 1$, by Lemma 5.3, we get

$$\begin{aligned} V(0) &= \lim_{x \to 0^+} \sup_{\overline{L} \in \Pi_{0,x}^L} V_{\overline{L}}(0) \\ &= \sup_{\overline{L} \in \Pi_{0,0}^L} V_{\overline{L}}(0) \\ &= V_{\overline{L}^0}(0). \end{aligned}$$

Suppose that $(x_0, x_1) \subset \mathcal{B}^*$ with $x_0 > 0$ and $x_0 \notin \mathcal{B}^*$; we now show that $x_0 \in \mathcal{A}^*$. If $V'(x_0) = 1$, the result is straightforward; we assume that this is not the case. We have that

$$\lim_{x \to x_0^+} \frac{V(x) - V(x_0)}{x - x_0} = 1.$$

Suppose that

$$\liminf_{x \to x_0^-} \frac{V(x) - V(x_0)}{x - x_0} = d_0 > 1,$$

by Definition 3.2, we have

$$\max\{1 - d, pd - (c+\beta)V(x_0) + \beta\mathcal{I}(V)(x_0))\} \geq 0$$

for all $d \in (1, d_0]$. Then,

$$pd - (c+\beta)V(x_0) + \beta\mathcal{I}(V)(x_0)) \geq 0$$

for all $d \in (1, d_0]$, and so $\Lambda(V)(x_0) \geq 0$. Therefore, by Lemma 5.1, we get $\Lambda(V)(x_0) = 0$. Now, if

$$\liminf_{x \to x_0^-} \frac{V(x) - V(x_0)}{x - x_0} = 1,$$

we show first that there exists a sequence $x_n \nearrow x_0$ such that $\lim_{n\to\infty} V'(x_n) = 1$. Indeed, take a sequence $h_n \searrow 0$ such that

$$\lim_{n\to\infty} \frac{V(x_0) - V(x_0 - h_n)}{h_n} = 1.$$

Define $a_n = (V(x_0) - V(x_0 - h_n)) / h_n - 1$, and let A_n be the set of all points $x \in [0, h_n]$ such that V is differentiable at x and $V'(x) \geq 1 + 2a_n$. By Proposition 1.3, we can assume that $a_n \geq 0$; if $a_n = 0$ for some n, we have that $V(x_0) - V(x) = x_0 - x$ for $x \in [x_0 - h_n, x_0]$, and so $V'(x_0) = 1$; if $a_n > 0$, using

$$a_n + 1 = \frac{\int_{A_n} V' + \int_{A_n^c} V'}{h_n} \geq \frac{|A_n|\,(1 + 2a_n) + (h_n - |A_n|)}{h_n},$$

we obtain that $|A_n| \leq \frac{h_n}{2}$ and so $|A_n^c| \geq \frac{h_n}{2}$. Then we can choose a sequence $x_n \nearrow x_0$ such that $1 \leq V'(x_n) < 1 + 2a_n$ and so $\lim_{n\to\infty} V'(x_n) = 1$. In the case that there exists a subsequence $x_{n_j} \nearrow x_0$ with $V'(x_{n_j}) > 1$, we have that $\Lambda(V)(x_{n_j}) = 0$ and, since $\mathcal{A}^*$ is a closed set, we obtain $\Lambda(V)(x_0) = 0$. In the case that $V'(x_n) = 1$ for all n and $\Lambda(V)(x_0) < 0$, let us show that there exists a point x_n close enough to x_0 in such a way that the function W_{x_n} introduced in Lemma 5.3 is a supersolution of (1.21) for all $x \in [x_n, x_0]$. By Lemma 5.3(a), we have

$$\begin{aligned}
\tilde{\mathcal{L}}_0(W_{x_n})(x) &= p - (c + \beta)W_{x_n}(x) + \beta\mathcal{I}(W_{x_n})(x) \\
&\leq p - (c + \beta)W_{x_n}(x) + \beta\mathcal{I}(V)(x_0) \\
&\leq \Lambda(V)(x_0) + (c + \beta)(V(x_0) - V(x_n)) \\
&< 0
\end{aligned}$$

for n large enough. Therefore, by Lemma 5.3(c), we conclude that $W_{x_n} = V$ in $[0, \hat{x}]$. This is a contradiction because $V'(x_0)$ does not exist. So we have that $x_0 \in \mathcal{A}^*$.

Let us prove now that $\mathcal{C}^*$ is right open. Given $x_0 \in \mathcal{C}^*$, we have that $\Lambda\,(V)\,(x_0) < 0$ and so, by Lemma 5.1, there exists $\delta > 0$ such that $\Lambda\,(V)\,(x) < 0$ in $(x_0, x_0 + \delta)$ and this implies that $(x_0, x_0 + \delta) \subset \mathcal{C}^* \cup \mathcal{B}^*$. If there were an element of $\mathcal{B}^*$ in $(x_0, x_0 + \delta)$, then the lower limit of the corresponding connected component would be a point of $\mathcal{A}^*$ in $(x_0, x_0 + \delta)$, but this is a contradiction.

Let us prove that there exists $b^* \geq 0$ such that $(b^*, \infty) \subset \mathcal{B}^*$. For each $y > 0$, consider the functions W_y introduced in Lemma 5.3; we will show that if y is large enough, then W_y is a viscosity supersolution of (1.21) for all $x \in (y, \infty)$ and the result follows from Lemma 5.3(b). Since $W_y' = 1$ in (y, ∞) we only need to show that $\tilde{\mathcal{L}}_0(W_y) \leq 0$ in (y, ∞). By Proposition 1.3, we have

$$\begin{aligned}\tilde{\mathcal{L}}_0(W_y)(x) &= p - (c+\beta)(V(y)+x-y) + \beta \int_{x-y}^{x} V(x-\alpha)dF(\alpha) \\ &\quad +\beta \int_0^{x-y} (V(y)+x-\alpha-y)\, dF(\alpha) \\ &\leq p - (c+\beta)(V(y)+x-y) + \beta \int_0^{x} (V(y)+x-\alpha-y)\, dF(\alpha) \\ &\leq p - c(V(y)+x-y)\end{aligned}$$

for all $x \geq y$. But

$$g_y(x) := p - c(V(y)+x-y)$$

is a decreasing function for each $y \geq 0$ and we have that $g_y(y) \leq 0$ for all the values y such that $V(y) \geq p/c$. Then, by Proposition 1.2, we get that $\tilde{\mathcal{L}}_0(W_y) \leq 0$ in (y,∞) for $y \geq p\beta/(c(c+\beta))$. Hence, the result follows for any $b^* > p\beta/(c(c+\beta))$ such that V is differentiable at b^*. □

In the next theorem, we prove that there exist optimal admissible strategies for all $x \geq 0$ and that they come from a band strategy.

Theorem 5.1. *The band strategy* $\pi(\mathcal{P}^*)$ *is optimal, that is* $V = V_{\mathcal{P}^*}$ *for all* $x \in \mathbf{R}_+$.

Proof. It is enough to see that V satisfies the conditions of Proposition 5.2 for the partition $\mathcal{P}^* = \{\mathcal{A}^*, \mathcal{B}^*, \ddot{\mathcal{C}}^*\}$. By Proposition 1.3, V is locally Lipschitz; by Definition 5.3, $V' = 1$ on $\mathcal{B}^*$ and $\Lambda(V) = 0$ in $\mathcal{A}^*$. We have that V is a viscosity solution of (1.21) and $V'(x) > 1$ at any x of $\mathcal{C}^*$ where V is differentiable, so V is an almost-everywhere solution of $\tilde{\mathcal{L}}_0 = 0$ in the connected components of C^*. □

In the following remarks, we give some results about the optimal value function and the optimal strategy for small and large surpluses.

Remark 5.3. Let us consider the simpler problem of finding the value function of the optimal barrier strategy (cf. Remark 5.2), that is

$$V_{bar}(x) = \sup\left\{V_{\mathcal{P}}(x) : \mathcal{P} = (\{a\},(a,\infty),[0,a)\}) \text{ with } a \geq 0\right\}.$$

From Proposition 5.3 and Theorem 5.1, we conclude that $V_{bar}(x) = V(x)$ for $x \leq \underline{a} := \min \mathcal{A}^*$ and that the best barrier strategy is $\mathcal{P} = (\{\underline{a}\},(\underline{a},\infty),[0,\underline{a})\}$. This does not imply in general that $V_{bar} = V$. On the other hand, the value of $V(0)$ is not known a priori, except in the case that $\underline{a} = 0$ where

$$V(0) = V_{bar}(0) = p/(c+\beta).$$

Remark 5.4. The existence of an optimal band strategy implies that there exists a surplus level $\overline{a} = \max \mathcal{A}^*$ such that the optimal dividend policy is to pay immediately the surplus exceeding $\overline{a}$. So under the optimal strategy, the controlled surplus process is smaller or equal than $\overline{a}$ after an eventual initial payment. We also have that $V(x) = V(\overline{a}) + x - \overline{a}$, and so V is not only locally Lipschitz, but Lipschitz.

5.3 Optimal Dividend Strategies with Reinsurance

In this section we introduce the concept of reinsurance band strategy and show that the optimal strategy exists and has this structure. Hereafter, we denote by $\mathcal{R}$ any of the reinsurance families introduced in Definition 2.1. Given a band partition $\mathcal{A}$, $\mathcal{B}$, and $\mathcal{C}$ as introduced in Definition 5.1 and a stationary reinsurance control ρ as introduced in Definition 2.2, we define the corresponding dividend band strategy with reinsurance.

Definition 5.4. Given a band partition $\mathcal{P} = (\mathcal{A}, \mathcal{B}, \mathcal{C})$, a stationary reinsurance control ρ, and an initial surplus $x \geq 0$, we define recursively an admissible strategy $\left(\overline{L}^x, \overline{R}^x\right) = (L_t^x, R_t^x)_{t\geq 0} \in \Pi_x^{L,R}$ as follows:

- In the case that $x \in \mathcal{A}$, then $L_t^x = p_{\rho^x} t$ and $R_t^x = \rho^x$ up to the first-claim arrival. Afterwards, follow the strategy corresponding to initial surplus $x - \rho^x(U_1)$ where U_1 is the size of the first claim.
- In the case that $x \in \mathcal{B}$, there exists an open interval $(x_0, x) \subset \mathcal{B}$ with $x_0 \in \mathcal{A}$; then $L_0^x = x - x_0$ and $R_0^x = \rho^x$. Afterwards, follow the strategy corresponding to initial surplus $x_0 \in \mathcal{A}$.
- In the case that $x \in \mathcal{C}$, there exists an open interval $(x, x_1) \subset \mathcal{C}$ with $x_1 \in \mathcal{A}$. Consider the unique surplus process X_t satisfying (2.5) with stationary reinsurance control ρ, and let us call $\tau_{\mathcal{C}}$ the first time that the surplus process exits $\mathcal{C}$. Then $L_t^x = 0$ and $R_t^x = \rho^{X_{t^-}}$ up to $\tau_{\mathcal{C}}$. Afterwards, follow the strategy corresponding to initial surplus $X_{\tau_{\mathcal{C}}}$.

  The family $\pi(\mathcal{P}, \rho) = \{\left(\overline{L}^x, \overline{R}^x\right) \in \Pi_x^{L,R}$ with $x \geq 0\}$ is called the *reinsurance band strategy* associated to $\mathcal{P}$ and ρ.

We define the value function $V_{\mathcal{P},\rho} : \mathbf{R}_+ \to \mathbf{R}$ of the reinsurance band strategy $\pi(\mathcal{P}, \rho)$ as $V_{\mathcal{P},\rho}(x) = V_{\overline{L}^x,\overline{R}^x}(x)$. There is a verification result for these value functions which is analogous to the one of Proposition 5.2 and the proof is similar. Given $R \in \mathcal{R}$, consider the operator Λ_R defined as

$$\Lambda_R(W)(x) = p_R - (c + \beta)W(x) + \beta \mathcal{I}_R(W)(x). \tag{5.12}$$

Proposition 5.5. *Consider a band partition $\mathcal{P} = (\mathcal{A}, \mathcal{B}, \mathcal{C})$ and a stationary reinsurance control ρ. Any Borel-measurable function W which is left continuous at the upper limit of the connected component of $\mathcal{C}$, it is right continuous at the lower limits of the connected components of $\mathcal{B}$, it has derivative equal to 1 on $\mathcal{B}$, it is an almost-everywhere solution of $\tilde{\mathcal{L}}_{\rho^x}(W) = 0$ in the connected components of $\mathcal{C}$, and it is a solution of $\Lambda_{\rho^x}(W) = 0$ in $\mathcal{A}$, should be the function $V_{\mathcal{P},\rho}$.*

We now show that the optimal value function is the value function of a band strategy $\pi(\mathcal{P}^*, \rho)$ associated to some reinsurance band partition $\mathcal{P}^*$ and some stationary reinsurance control ρ. We define the reinsurance band partition $\mathcal{P}^*$ in a similar way to Definition 5.3 based on the optimal value function V defined in (2.16).

Definition 5.5. Let us define $\mathcal{P}^* = (\mathcal{A}^*, \mathcal{B}^*, \mathcal{C}^*)$

- $\mathcal{A}^* = \{x \in \mathbf{R}_+$ such that $\sup_{R\in\mathcal{R}} \Lambda_R(V)(x) = 0\}$.
- $\mathcal{B}^* = \{x \in (0, \infty)$ such that $V'(x) = 1$ and $\sup_{R\in\mathcal{R}} \Lambda_R(V)(x) < 0\}$.
- $\mathcal{C}^* = (\mathcal{A}^* \cup \mathcal{B}^*)^c$.

Remark 5.5. As in the case without reinsurance (see Proposition 5.3), it can be proved that if $\hat{x} \in \mathcal{A}^* \cup \mathcal{B}^*$, then for any $x \leq \hat{x}$ we have

$$V(x) = \sup_{(\overline{L},\overline{R})\in\Pi_{x,\hat{x}}^{L,R}} V_{\overline{L},\overline{R}}(x),$$

where

$$\Pi_{x,\hat{x}}^{L,R} = \{(\overline{L}, \overline{R}) \in \Pi_x^{L,R} \text{ such that } X_t^{\overline{L},\overline{R}} \leq \hat{x} \text{ for all } t \geq 0\}.$$

Lemma 5.4. *Since V is a viscosity solution of (2.19), then $V' > 1$ and $\sup_{R\in\mathcal{R}} \tilde{\mathcal{L}}_R(V)(x) = 0$ at all the points of $\mathcal{C}^*$ where V is differentiable.*

As in Lemma 5.1, we have the following result.

Lemma 5.5. *The function $\sup_{R\in\mathcal{R}} \Lambda_R(V)$ is right continuous and upper semicontinuous and satisfies $\sup_{R\in\mathcal{R}} \Lambda_R(V) \leq 0$.*

The details of the proofs can be found in Proposition 7.4 and Corollary 7.5 of [9].

Definition 5.6. Let us define the function

$$\hat{V}(x) = \inf_{R\in\mathcal{R}} \left\{ \frac{(c+\beta)V(x) - \beta\mathcal{I}_R(V)(x)}{p_R} \right\}$$

and the function

$$\hat{I}_R(V)(x) = p_R\hat{V}(x) - (c+\beta)V(x) + \beta\mathcal{I}_R(V)(x)$$

for any $R \in \mathcal{R}$.

Lemma 5.6. *$\hat{V}$ is well defined and Borel measurable, $\hat{V} \geq 1$, and*

$$\sup_{R\in\mathcal{R}} \left\{ \hat{I}_R(V)(x) \right\} = 0.$$

Moreover, $\hat{V} > 1$ in $\mathcal{C}^*$, $\hat{V} = 1$ in $\mathcal{A}^*$, and $\hat{V}(x) = V'(x)$ at all the points of $x \in \mathcal{C}^*$ where V is differentiable (i.e., a.e. in $\mathcal{C}^*$).

Proof. $\hat{V}$ is well defined because for any $R \in \mathcal{R}$,

$$0 \leq \hat{V}(x) \leq \frac{(c+\beta)V(x) - \beta\mathcal{I}_R(V)(x)}{p_R} \leq \frac{(c+\beta)V(x)}{p}.$$

By definition, we have that $\sup_{R\in\mathcal{R}}\left\{\hat{l}_R(V)(x)\right\} = 0$ for all $x \geq 0$. By Proposition 5.5, we get $\hat{V} \geq 1$. If $\hat{V}(x) = 1$, then $\sup_{R\in\mathcal{R}} \Lambda_R(V)(x) = 0$ and so $x \in \mathcal{A}^*$; therefore $\hat{V}(x) > 1$ for all $x \in \mathcal{C}^* \cup \mathcal{B}^*$. If $x \in \mathcal{C}^*$ and $V'(x)$ exists, then $V'(x) > 1$ and so $\sup_{R\in\mathcal{R}} \tilde{\mathcal{L}}_R(V)(x) = 0$, because V is a viscosity solution of (2.19); then $\hat{V}(x) = V'(x)$ by Definition 5.6. □

Proposition 5.6. *There exists a reinsurance policy $R_x^* \in \mathcal{R}$ such that*

$$\sup_{R\in\mathcal{R}}\left\{\hat{l}_R(V)(x)\right\} = \hat{l}_{R_x^*}(V)(x) = 0$$

with $p_{R_x^*} \geq cpV(0)/((c+\beta)V(x))$ for any $x \in \mathcal{A}^* \cup \mathcal{C}^*$.

Proof. The result is straightforward for $\mathcal{R}_F$. Given $x \geq 0$, we first prove that in each of the cases, there exists a retained loss function R_x^* where the supremum is attained, and then we prove that $p_{R_x^*}$ is positive.

Family of all the proportional retained loss functions:

Given any $b \in [0,1]$, we denote by R_b the proportional reinsurance policy defined as $R_b(\alpha) = b\alpha$. We define for any $x \geq 0$ the function $g_P(V,x,\cdot) : [0,1] \to \mathbf{R}$ as

$$g_P(V,x,b) = p_{R_b}\hat{V}(x) + \beta\mathcal{I}_{R_b}(V)(x). \tag{5.13}$$

Note that $g_P(V,x,b) = \hat{l}_{R_b}(V)(x) + (c+\beta)V(x)$. Then we can write

$$g_P(V,x,b) = \beta E(U_i)\hat{V}(x)\left((1+\eta_1)b - (\eta_1 - \eta)\right) + \beta\int_0^\infty V(x-b\alpha)\,dF(\alpha).$$

Since

$$g_P(V,x,b^-) = g_P(V,x,b) = g_P(V,x,b^+) + \beta V(0)\left(F\left(\frac{x}{b}\right) - F\left(\left(\frac{x}{b}\right)^-\right)\right)$$

for $b > 0$ and

$$g_P(V,x,0) = g_P(V,x,0^+) = -\beta E(U_i)\hat{V}(x)(\eta_1 - \eta),$$

we have that $g_P(V,x,\cdot)$ is a well defined, left-continuous function with negative jumps. So there exists at least one value $b \in [0,1]$ where the maximum of $g_P(V,x,\cdot)$ is attained. Let us define

$$b^*(x) = \max(\arg\max_{b\in[0,1]} g_P(V,x,b)). \tag{5.14}$$

Defining $R_x^*(\alpha) = b^*(x)\alpha$, we have that $R_x^* = \arg\max_{R_b}\left\{\hat{l}_{R_b}(V)(x)\right\}$.

Family of all the excess-of-loss retained loss functions:

Given $x \geq 0$, we consider the excess-of-loss reinsurance policy $R_a(\alpha) = a \wedge \alpha$ for any $a \in [0, \infty]$ and define the function $g_{XL}(V, x, \cdot) : [0, \infty] \to \mathbf{R}$ as

$$g_{XL}(V, x, a) = p_{R_a} \hat{V}(x) + \beta \mathcal{I}_{R_a}(V)(x).$$

We have that

$$p_{R_a} = \beta(1+\eta_1)\left(\int_0^a \alpha dF(\alpha) + a\,(1 - F(a))\right) - \beta(\eta_1 - \eta)E(U_i)$$

and

$$\int_0^\infty V(x - R_a(\alpha))\, dF(\alpha) = \int_0^a V(x-\alpha)dF(\alpha) + V(x-a)(1 - F(a).$$

Therefore,

$$\begin{aligned} g_{XL}(V, x, a) = \beta \textstyle\int_0^a & \left((1+\eta_1)\hat{V}(x)\alpha + V(x-\alpha)\right) dF(\alpha) \\ & + \beta(1 - F(a))\left(\hat{V}(x)(1+\eta_1)a + V(x-a)\right) \\ & - \beta(\eta_1 - \eta)E(U_i)\hat{V}(x) \end{aligned} \tag{5.15}$$

for $a \in \mathbf{R}_+$ and

$$\begin{aligned} g_{XL}(V, x, \infty) = \beta \textstyle\int_0^\infty & \left((1+\eta_1)\hat{V}(x)\alpha + V(x-\alpha)\right) dF(\alpha) \\ & - \beta(\eta_1 - \eta)E(U_i)\hat{V}(x). \end{aligned}$$

Note that $g_{XL}(V, x, \cdot)$ is a continuous function in $[0, x)$,

$$g_{XL}(V, x, x^-) = g_{XL}(V, x, x) > g_{XL}(V, x, x^+)$$

and $g_{XL}(V, x, a) \leq g_{XL}(V, x, \infty)$ for $a > x$, so the maximum of $g_{XL}(V, x, \cdot)$ is attained at least at a point of $[0, x] \cup \{\infty\}$; we define

$$a^*(x) = \max(\arg\max_{a \in [0,x] \cup \{\infty\}} g_{XL}(V, x, a)). \tag{5.16}$$

Defining $R_x^*(\alpha) = a^*(x) \wedge \alpha$, we have that $R_x^* = \arg\max_{R_a} \left\{ \hat{I}_{R_a}(V)(x) \right\}$.

Family of all the retained loss functions:

Given $x \geq 0$, let us characterize the reinsurance policy that maximizes the function

$$\begin{aligned} g_R(V, x) = \beta \textstyle\int_0^\infty & \left((1+\eta_1)\hat{V}(x)R(\alpha) + V(x - R(\alpha))\right) dF(\alpha) \\ & - \beta(\eta_1 - \eta)\hat{V}(x)E(U_i) \end{aligned} \tag{5.17}$$

among all the F-measurable functions $R : \mathbf{R}_+ \to \mathbf{R}_+$ such that $0 \le R(\alpha) \le \alpha$. Note that $g_R(V, x) = \hat{I}_R(V)(x) + (c + \beta)V(x)$. In order to maximize $g_R(V, x)$ among all the possible retained loss functions, we find the maximum of the integrand for each α, i.e., the maximum for each $x \ge 0$ is attained at the retained loss functions defined as

$$R_x^*(\alpha) = \max\left(\arg\max_{s\in[0,\alpha]}\left((1+\eta_1)\hat{V}(x)s + V(x-s)\right)\right). \tag{5.18}$$

This maximum exists because $(1+\eta_1)\hat{V}(x)s + V(x-s)$ is a continuous function except for a negative jump at x where it is left continuous.

Finally, let us show now that $p_{R_x^*}$ is positive in all the cases. If $x \in \mathcal{A}^* \cup \mathcal{C}^*$, since V is increasing, we have

$$0 = \hat{I}_{R_x^*}(V)(x) \le p_{R_x^*}\hat{V}(x) - cV(x) \le p_{R_x^*}\hat{V}(x) - cV(0),$$

and so, by Lemma 5.6,

$$p_{R_x^*} \ge \frac{cV(0)}{\hat{V}(x)} \ge \frac{cpV(0)}{(c+\beta)V(x)} > 0.$$

□

As in Sect. 5.2, the sets introduced in Definition 5.5 have the suitable properties.

Proposition 5.7. *$\mathcal{P}^* = (\mathcal{A}^*, \mathcal{B}^*, \mathcal{C}^*)$ is a band partition.*

The proof is similar to the one of Proposition 5.4. See the details in Proposition 8.2 of [9].

We define the function $\rho : \mathbf{R}_+ \times \mathbf{R}_+ \to \mathbf{R}$ as

$$\rho^x(\alpha) = \begin{cases} R_x^*(\alpha) \text{ if } x \in \mathcal{A}^* \cup \mathcal{C}^* \\ R_0(\alpha) \text{ if } x \in \mathcal{B}^*, \end{cases} \tag{5.19}$$

where R_x^* is defined in Proposition 5.6 and $R_0 \in \mathcal{R}$ is any retained loss function (the choice of R_0 is irrelevant).

Proposition 5.8. *The function ρ defined in (5.19) is a stationary reinsurance control. Moreover, there exists $p_0 > 0$ such that $p_{\rho^x} \ge p_0$ for all $x \ge 0$.*

Proof. Let us prove first that ρ is Borel measurable. This result is straightforward for the family $\mathcal{R}_F$. For the other three families, $\rho^x = R_0$ for $x \in \mathcal{B}^*$ and, for $x \in \mathcal{A}^* \cup \mathcal{C}^*$, $\rho^x(\alpha) = b^*(x)\alpha$ in the case $\mathcal{R} = \mathcal{R}_P$, $\rho^x(\alpha) = \alpha \wedge a^*(x)$ in the case $\mathcal{R} = \mathcal{R}_{XL}$, and $\rho^x = R_x^*$ in the case $\mathcal{R} = \mathcal{R}_A$. The Borel-measurable functions $b^*(x)$, $a^*(x)$, and R_x^* are defined in (5.14), (5.16), and (5.18), respectively. Since, by Proposition 5.7, there exists $b \ge 0$ such that $\mathcal{A}^* \cup \mathcal{C}^* \subset [0, b]$, we have that $0 < V(x) \le V(b)$ for all $x \in \mathcal{A}^* \cup \mathcal{C}^*$ and so

$$p_{\rho^x} \geq p_0 := \min\{\frac{cpV(0)}{(c+\beta)V(b)}, p_{R_0}\} > 0.$$

□

The following theorem is a direct consequence of Propositions 5.5, 5.7, and 5.8. The proof is similar to the one of Theorem 5.1.

Theorem 5.2. *The reinsurance band strategy $\pi(\mathcal{P}^*, \rho)$ associated to $\mathcal{P}^*$ introduced in Definition 5.5 and ρ defined in (5.19) is optimal, that is, $V = V_{\mathcal{P}^*,\rho}$ for all $x \in \mathbf{R}_+$.*

The results for small and large surpluses of Remarks 5.3 and 5.4 hold for the case of reinsurance in the families $\mathcal{R}_P$, $\mathcal{R}_{XL}$, and $\mathcal{R}_A$. In the family $\mathcal{R}_F$, the remarks also hold, except that if $\min \mathcal{A}^* = 0$, then

$$V(0) = \frac{\max_{R \in \mathcal{R}_F} \{p_R\}}{c+\beta}.$$

5.4 Optimal Dividend Strategies with Investments

In this section we modify the idea of band strategies introduced in Sect. 5.1 to study the problem of dividends and investment. A key property of the optimal dividend strategy in Sects. 5.2 and 5.3 is the existence of certain levels a of surplus where the best policy is to pay as dividends the amount exceeding a if the surplus $x \in (a, a+\varepsilon)$, to pay no dividends if the surplus $x \in (a-\varepsilon, a)$, and to pay the incoming premium rate as dividends if the surplus coincides with a; when this point a is reached, the surplus remains constant up to the occurrence of the next claim. Note that in the case of investment control, the third condition does not hold in general: the only way that the surplus remains constant at level a up to the occurrence of the next claim is when the fraction of surplus invested in stocks is zero. However, we will show that there are actually levels a where the optimal dividend strategy is to pay as dividends the amount exceeding a if the surplus $x \in (a, a+\varepsilon)$, to pay no dividends if the surplus $x \in (a-\varepsilon, a)$, but the best investment strategy at a is positive. In order to overcome this problem, we introduce the notion of limit of stationary band strategies and show that the optimal strategy could be written as such a limit.

Let us first introduce the concept of limit band strategy in the simplest case where there is just one band (barrier). Roughly speaking, a dividend policy is called barrier with level a when all excess surplus above a is paid out immediately as dividends, but there are no dividend payments when surplus is less than a. The issue is to determine what to do when the surplus is a. We define a limit barrier strategy as an explicit limit of stationary barrier strategies and find its value function.

Let us assume from now on that $\Gamma = [0, \hat{\gamma}]$ for some investment constraint $\hat{\gamma}$ [recall that we assumed that $\hat{\gamma} < c/r$ for the dividend case in (2.34)]. Take any positive and Lipschitz function $g : [0, a] \to \Gamma$ satisfying that there exists $\delta > 0$ such that

$$g(x) = \hat{\gamma} \text{ for } x \leq \delta, \tag{5.20}$$

and let us extend g as $g(x) = g(a)$ for $x > a$. Then g is a stationary investment control as defined in Definition 2.3. We assume that $g = \hat{\gamma}$ for small values to avoid technical issues; we will see later that there exists an optimal stationary investment control and that it satisfies this condition.

Definition 5.7. Given a stationary investment control g satisfying (5.20) and $0 < u < a$, we define recursively the admissible strategy $\pi_x^{(a,u)} = (\overline{L}, \overline{\gamma}) \in \Pi_x^{L,\gamma}$ for any initial surplus $x \geq 0$ as follows:

- In the case that $x < a$, consider the processes X_t^g and $\overline{\gamma}^g$ defined in Remark 2.3; let us call τ the ruin time of the process X_t^g and τ_a the first time that the surplus X_t^g reaches a; take $L_t = 0$ and $\gamma_t = \gamma_t^g$ up to $\tau^* = \tau_a \wedge \tau$. If $\tau^* = \tau_a$, follow the dividend and investment strategy corresponding to initial surplus a.
- If $x = a > 0$, pay immediately u as dividends and follow the strategy $\pi_{a-u}^{(a,u)} \in \Pi_{a-u}^{L,\gamma}$.
- If $x > a$, pay immediately $x - a$ as dividends and follow the strategy $\pi_a^{(a,u)} \in \Pi_a^{L,\gamma}$.

The family $(\pi_x^{(a,u)})_{x \geq 0}$ is called a *stationary barrier strategy*.

In the extreme case that $a = 0$, the barrier dividend policy is to pay out immediately all the surplus and then to pay the incoming premium p as dividends up to the arrival time of the first claim (i.e., the ruin time).

Given the stationary barrier strategy $(\pi_x^{(a,u)})_{x \geq 0}$, we define its value function $V^{(a,u)}$ as $V^{(a,u)}(x) = V_{\pi_x^{(a,u)}}(x)$ for $x \geq 0$. In the next proposition, we write $V^{(a,u)}$ in terms of the unique twice continuously differentiable solution W of the equation $\tilde{\mathcal{L}}_{g(x)}(W)(x) = 0$ with $W(0) = 1$. The existence, uniqueness, and regularity of W follow from a fixed-point argument; the proofs are similar to the ones given in Sects. 3 and 4 of [11].

Proposition 5.9. *Given* $0 < u < a$ *, we have*

$$V^{(a,u)}(x) = \begin{cases} \frac{W(x)}{(W(a)-W(a-u))/u} & \text{if } 0 \leq x < a \\ \frac{W(a)}{(W(a)-W(a-u))/u} + (x-a) & \text{if } x \geq a, \end{cases}$$

and $V^{(a,u)}(x) = p/(c+\beta) + x$ *if* $a = 0$ *(here* $V^{(a,u)}$ *does not depend on* u*).*

Proof. If $a = 0$, the result is straightforward; let us consider the case $a > 0$. We extend the definition of W as $W = 0$ in $(-\infty, 0)$. Consider an initial surplus

$0 \le x < a$, and let $\overline{L}, X_t^g$, τ^*, τ_a and τ be as in Definition 5.7. The function W is twice continuously differentiable and by Proposition 2.7, $\tau^*$is finite. From Proposition 2.13, we have $W(x) = E_x(W(X_{\tau^*}^g)e^{-c\tau^*})$. Then we can write

$$W(x) = E_x(W(X_{\tau^*}^g)e^{-c\tau^*} I_{\{\tau^*=\tau_a\}}) = W(a)E_x(e^{-c\tau^*} I_{\{\tau^*=\tau_a\}})$$

and so $E_x(e^{-c\tau^*} I_{\{\tau^*=\tau_a\}}) = W(x)/W(a)$. On the other hand, from Definition 5.7,

$$V^{(a,u)}(x) = E_x\left(e^{-c\tau^*} V^{(a,u)}(X_{\tau^*}^g)\right) = V^{(a,u)}(a)E_x(e^{-c\tau^*} I_{\{\tau^*=\tau_a\}})$$

and the result follows from $V^{(a,u)}(a) = V^{(a,u)}(a-u) + u$. □

We say that a sequence of stationary strategies converges if the sequence of their value functions converges pointwise. This limit of value functions is called *the value function of the limit strategy*. By Proposition 5.9, we obtain

$$\lim_{u\to 0} V^{(a,u)}(x) = \begin{cases} W(x)/W'(a) & \text{if } 0 \le x < a \\ W(a)/W'(a) + (x-a) & \text{if } x \ge a. \end{cases} \tag{5.21}$$

So, we can define the *limit barrier strategy with barrier a* as $(\tilde{\pi}_x^a)_{x\ge 0} = (\lim_{u\to 0} \pi_x^{(a,u)})_{x\ge 0}$ (in the special case $a = 0$, there is no need to take this limit).

Our next step is to define the limit band strategy associated to botha band partition and a stationary investment control. The definition of band partition for the investment and dividend payments problem differs slightly from the definition used in the previous sections.

Definition 5.8. We say that $\mathcal{P} = (\mathcal{A}, \mathcal{B}, \mathcal{C})$ is an *(investment) band partition* if $\mathcal{A}, \mathcal{B}$, and $\mathcal{C}$ are disjoint sets and satisfy the following properties: $\mathbf{R}_+ = \mathcal{A} \cup \mathcal{B} \cup \mathcal{C}$, $\mathcal{C}$ is an open set in $\mathbf{R}_+$, $\mathcal{B}$ is a disjoint union of intervals that are left open and right closed, the lower boundary of any connected component of $\mathcal{B}$ belongs to $\mathcal{A}$, and there exists $b \in \mathcal{A}$ such that $(b, \infty) \subset \mathcal{B}$. We also require the condition that there is not isolated points of $\mathcal{A}$ in $\mathcal{C}$: that is, if $(x-\vartheta, x) \cup (x, x+\vartheta) \subset \mathcal{C}$ for some $\vartheta > 0$, then $x \in \mathcal{C}$.

Remark 5.6. This definition implies that the upper boundary of any connected component of $\mathcal{C}$ belongs to $\mathcal{A}$, that both $\mathcal{A}$ and $\mathcal{C}$ are bounded sets, and that $0 \in \mathcal{A} \cup \mathcal{C}$. The sets $\mathcal{A}$ and $\mathcal{B}$ are nonempty, but $\mathcal{C}$ could be empty [in this case $\mathcal{A} = \{0\}$ and $\mathcal{B} = (0, \infty)$].

A band partition with $\mathcal{A} = \{a\}$ corresponds to a barrier partition. Before considering a general band partition (where $\#\mathcal{A}$ can be infinite), we define limit band strategies for finite band partitions.

Definition 5.9. An investment band partition is *finite* if $\#\mathcal{A}$ is finite. A finite band partition with $\#\mathcal{A} = n$ could be written as $\mathcal{A} = \{a_1, .., a_n\}$, $\mathcal{B} = \bigcup_{i=1}^{n-1}(a_i, b_i] \cup (a_n, \infty)$ and $\mathcal{C} = [0, a_1) \cup \bigcup_{i=1}^{n-1}(b_i, a_{i+1})$ with $0 \le a_1 < b_1 < a_2 < b_2 < \cdots < a_n$.

Let us define a stationary finite band strategy associated to a finite band partition and a stationary investment control.

Definition 5.10. Given a finite band partition $\mathcal{P}$ as defined in (5.9), a stationary investment control g satisfying (5.20) and $u > 0$ small enough such that $0 \le a_1 - u \in \mathcal{C}$ for $a_1 > 0$ and $b_i < a_{i+1} - u \in \mathcal{C}$ for $i = 2, \ldots, n-1$; we define recursively the admissible strategy $\pi_x(\mathcal{P}, g, u) = (\overline{L}, \overline{\gamma}) \in \Pi_x^{L,\gamma}$ for any initial surplus $x \ge 0$ as follows:

- In the case that $x \in \mathcal{C}$, consider the process X_t^g and $\overline{\gamma}^g$ defined in Remark 2.3; take $L_t = 0$ and $\gamma_t = \gamma_t^g$ up to $\tau_{\mathcal{C}}$, where $\tau_{\mathcal{C}}$ is the exit time of the process X_t^g of the set $\mathcal{C}$. Afterwards, if $\tau_{\mathcal{C}} < \tau$ (where τ is the ruin time of the process), follow the dividend and investment strategy corresponding to initial surplus $X_{\tau_{\mathcal{C}}}^g \in \mathcal{A} \cup \mathcal{B}$.
- If $x = a_i \in \mathcal{A}$ and $a_i > 0$, pay immediately u as dividends and follow the strategy $\pi_{a_i - u}(\mathcal{P}, g, u) \in \Pi_{a_i - u}^{L,\gamma}$, where $a_i - u \in \mathcal{C}$.
- If $x = a_1 = 0 \in \mathcal{A}$, pay the incoming premium p as dividends up to the arrival time of the first claim (which is the ruin time).
- If $x \in (a_i, b_i] \subset \mathcal{B}$, $i = 1, \ldots, n-1$, pay immediately $x - a_i$ as dividends and follow the strategy $\pi_{a_i}(\mathcal{P}, g, u) \in \Pi_{a_i}^{L,\gamma}$. Similarly, if $x > a_n$, pay immediately the surplus $x - a_n$ as dividends and follow the strategy $\pi_{a_n}(\mathcal{P}, g, u) \in \Pi_{a_n}^{L,\gamma}$.

As in the case of barrier strategies, given a stationary finite band strategy $(\pi_x(\mathcal{P}, g, u))_{x \ge 0}$, we can define its value function $V_{(\mathcal{P},g,u)}$ as $V_{(\mathcal{P},g,u)}(x) = V_{\pi_x(\mathcal{P},g,u)}(x)$ for $x \ge 0$. The case of limit barrier strategy suggests that $\lim_{u \to 0} V_{(\mathcal{P},g,u)} = \overline{W}$, where $\overline{W}$ can be written in terms of solutions of the equation $\tilde{\mathcal{L}}_{g(x)} = 0$. Indeed, $\overline{W}$ is the unique Lipschitz function which satisfies:

- $\overline{W}(0) = p/(c + \beta)$ if $a_1 = 0$.
- $\overline{W}$ is the unique twice continuously differentiable solution of $\tilde{\mathcal{L}}_{g(x)} = 0$ with boundary condition $\overline{W}'(a_1) = 1$ in $[0, a_1)$ if $a_1 > 0$.
- $\overline{W}(x) = \overline{W}(a_i) + x - a_i$ if $x \in (a_i, b_i]$ (for $i = 1, \ldots, n-1$).
- $\overline{W}(x) = \overline{W}(a_n) + x - a_n$ for $x > a_n$.
- $\overline{W}$ is the unique twice continuously differentiable solution of $\tilde{\mathcal{L}}_{g(x)} = 0$ in (b_i, a_{i+1}) with boundary conditions $\overline{W}(b_i^+) = \overline{W}(b_i)$ and $\overline{W}'(a_{i+1}) = 1$ (for $i = 1, \ldots, n-1$).

We define the limit band strategy as $\tilde{\pi}(\mathcal{P}, g) = (\tilde{\pi}_x(\mathcal{P}, g))_{x \ge 0} = (\lim_{u \to 0} \pi_x(\mathcal{P}, g, u))_{x \ge 0}$ and $V_{\tilde{\pi}(\mathcal{P},g)} = \overline{W}$.

Finally, let us consider the general case of an infinite band partition $\mathcal{P} = (\mathcal{A}, \mathcal{B}, \mathcal{C})$ as introduced in Definition 5.8. We define the limit band strategy as a double limit: on one hand, we approximate the partition $\mathcal{P}$ by a finite partition $\mathcal{P}_\delta = (\mathcal{A}_\delta, \mathcal{B}_\delta, \mathcal{C}_\delta)$ and the limit band strategy $(\tilde{\pi}_x(\mathcal{P}_\delta, g))_{x \ge 0}$ by the stationary finite band strategy $(\pi_x(\mathcal{P}_\delta, g, u))_{x \ge 0}$.

Let us explain how we construct finite partitions $\mathcal{P}_\delta$ which approximate $\mathcal{P}$. Given $\delta > 0$ small enough, we define $\mathcal{A}_\delta$ as the set of points $a \in \mathcal{A}$ which are upper boundaries of the connected component of $\mathcal{C}$ and satisfy that there is no other point of $\mathcal{A}$ in $(a - \delta, a)$. We also require that, if $0 \in \mathcal{A}$, then $0 \in \mathcal{A}_\delta$. Note that $\#\mathcal{A}_\delta \leq \hat{x}/\delta + 1$ because $\mathcal{A} \subset [0, \hat{x}]$. If either $0 \in \mathcal{A}$ or there exists a connected component of $\mathcal{C}$ with length larger than δ, then $\mathcal{A}_\delta$ is nonempty. The only way that $\mathcal{C}$ could be empty is when $\mathcal{A} = \{0\}$ and $\mathcal{B} = (0, \infty)$, so $\mathcal{A}_\delta$ is nonempty for δ small enough. We define $\mathcal{C}_\delta$ as the (finite) union of the connected component of $\mathcal{C}$ whose upper boundaries belong to $\mathcal{A}_\delta$ and $\mathcal{B}_\delta$ as $\mathbf{R}_+ \backslash (\mathcal{A}_\delta \cup \mathcal{C}_\delta)$. We have that $\mathcal{P}_\delta$ is a finite band partition as introduced in Definition 5.9. Let $\varsigma > 0$ be the minimum of the length of the connected component of $\mathcal{C}_\delta$.

Given a stationary investment control g satisfying (5.20), a number $\delta > 0$ small enough, and $u \in (0, \varsigma)$, we consider the stationary finite band strategy $(\pi_x(\mathcal{P}_\delta, g, u))_{x \geq 0}$ and define the limit band strategy $(\pi_x(\mathcal{P}, g))_{x \geq 0}$ as $\pi_x(\mathcal{P}, g) = \lim_{(\delta,u)\to 0} \pi_x(\mathcal{P}_\delta, g, u)$ (if $\lim_{(\delta,u)\to 0} V_{(\mathcal{P}_\delta, g, u)}$ exists). The values of the stationary investment control g outside the set $\mathcal{C}$ are irrelevant so it is enough to define g on $\mathcal{C}$.

Remark 5.7. If $\mathcal{P} = (\mathcal{A}, \mathcal{B}, \mathcal{C})$ is a finite band partition, then $\mathcal{P}_\delta = \mathcal{P}$ for δ small enough; and so $\pi_x(\mathcal{P}, g) = \lim_{u\to 0} \pi_x(\mathcal{P}, g, u)$.

The main result of the section is to show that there exists an optimal band partition $\mathcal{P}^*$ and an optimal stationary investment control g^* satisfying (5.20) such that the $\lim_{(\delta,u)\to 0} V_{(\mathcal{P}_\delta, g, u)}$ is the optimal value function V defined in (2.33); and so, there exists an optimal limit band strategy. Similarly to the previous sections, the optimal investment band partition $\mathcal{P}^*$ and the optimal stationary investment control g^* can be obtained from the optimal value function V.

We will not prove the results in the remainder of the section because the main ideas of the proofs are similar to the ones in the case without investment. For detailed proofs of all these results, see Sect. 8 of [11]. In this paper, the problem is solved for the case $\Gamma = [0, 1]$, but all results hold for the more general case $\Gamma = [0, \hat{\gamma}]$ considered here.

In order to define $\mathcal{P}^*$, we first introduce some auxiliary sets based on the optimal value function V. Let us consider the operator

$$\Lambda_D(V)(x) = (p + rx\hat{\gamma}) - (c + \beta)V(x) + \beta\mathcal{I}(V)(x), \qquad (5.22)$$

and the sets

- $\mathcal{A}_1 = \{x \in \mathbf{R}_+ \text{ such that } V'(x) = 1 \text{ and } \Lambda_D(V)(x) = 0\}$,
- $\mathcal{B}_1 = \{x \in (0, \infty) \text{ such that } V'(x) = 1 \text{ and } \Lambda_D(V)(x) < 0\}$,
- $\mathcal{C}_1 = \mathbf{R}_+ - (\mathcal{A}_1 \cup \mathcal{B}_1)$.

We need to modify slightly the sets defined above removing some annoying points from the set $\mathcal{A}_1$.

Definition 5.11. We define the sets $\mathcal{A}^*$, $\mathcal{B}^*$, and $\mathcal{C}^*$ as

- $\mathcal{B}^* = \mathcal{B}_1 \cup \{a \in \mathcal{A}_1 : (a - \vartheta, a) \subset \mathcal{A}_1 \cup \mathcal{B}_1$ for some $\vartheta > 0\}$,
- $\mathcal{C}^* = \mathcal{C}_1 \cup \{a \in \mathcal{A}_1 : (a - \vartheta, a) \cup (a, a + \vartheta) \subset \mathcal{C}_1$ for some $\vartheta > 0\}$,
- $\mathcal{A}^* = \mathbf{R}_+ - (\mathcal{C}^* \cup \mathcal{B}^*)$.

Like in Proposition 5.4, we obtain the following result (cf. Proposition 8.5 of [11]).

Proposition 5.10. *$\mathcal{P}^* = (\mathcal{A}^*, \mathcal{B}^*, \mathcal{C}^*)$ is a band partition as defined in Definition 5.8.*

Remark 5.8. As in the other dividend control problems, it can be proved that if $\hat{x} \in \mathcal{A}^* \cup \mathcal{B}^*$, then

$$V(x) = \sup\nolimits_{(\overline{L},\overline{\gamma}) \in \Pi_{x,\hat{x}}^{L,\gamma}} V_{\overline{L},\overline{\gamma}}(x),$$

where

$$\Pi_{x,\hat{x}}^{L,\gamma} = \{(\overline{L},\overline{\gamma}) \in \Pi_x^{L,\gamma} \text{ such that } X_t^{\overline{L},\overline{\gamma}} \leq \hat{x} \text{ for all } t \geq 0\}.$$

The proof can be found in Proposition 8.1 in [11].

Using a fixed-point argument, and taking into consideration the regularity of the claim-size distribution F, we have the following regularity result on the optimal value function. See Propositions 6.3, 8.6, and 8.8 of [11].

Proposition 5.11. *V is continuously differentiable in $\mathbf{R}_+$; it is twice continuously differentiable with bounded second derivative in $\mathcal{C}^*$ and in the interior of $\mathcal{B}^*$ (the second derivative could not exist at the remaining points). Moreover, V is a classical solution of the equation $\sup_{\gamma\in\Gamma} \tilde{\mathcal{L}}_\gamma(V)(x) = 0$ for $x \in \mathcal{C}^*$ and*

$$\arg\max_{\gamma\in\Gamma} \tilde{\mathcal{L}}_\gamma(V)(x) = \min\{\hat{\gamma}, 2\tfrac{(c+\beta)V(x) - \beta\mathcal{I}(V)(x) - pV'(x)}{r\hat{\gamma}xV'(x)}\}$$

for $x \in \mathcal{C}^* \backslash \{0\}$. If $0 \in \mathcal{C}^*$, then $\arg\max_{\gamma\in\Gamma} \tilde{\mathcal{L}}_\gamma(V)(x) = \hat{\gamma}$ for small values of x.

We use this last formula to define the investment control g^* as

$$g^*(x) = \min\{\hat{\gamma}, 2\frac{(c+\beta)V(x) - \beta\mathcal{I}(V)(x) - pV'(x)}{r\hat{\gamma}xV'(x)}\} \tag{5.23}$$

for $x \in \mathcal{C}^* \backslash \{0\}$ and $g^*(0) = \hat{\gamma}$ if $0 \in \mathcal{C}^*$. Note that g^* is Lipschitz in $\mathcal{C}^*$ because $\mathcal{C}^* \subset [0, \hat{x}]$, $V' > 1$ and V is twice continuously differentiable in $\mathcal{C}^*$ with bounded second derivative in $\mathcal{C}^*$. So g^* is a stationary investment control satisfying (5.20).

Finally, we have the main result. This result corresponds to Theorem 8.11 of [11].

Theorem 5.3. *Consider the investment band partition* $\mathcal{P}^*$ *and the stationary investment control* g^* *defined in Definition 5.11 and (5.23), respectively. The optimal value function* V *defined in (2.33) satisfies* $V = \lim_{(\delta,u)\to 0} V_{(\mathcal{P}^*_\delta, g^*, u)}$, *and so the limit band strategy associated with* $\mathcal{P}^*$ *and* g^* *is optimal.*

5.5 Optimal Reinsurance Control for Survival Probability

Let us analyze the existence of the optimal stationary reinsurance control ρ in the problem of maximizing the survival probability with reinsurance defined in (2.9). That is the existence of a Borel-measurable function $\rho : \mathbf{R}_+ \times \mathbf{R}_+ \to \mathbf{R}$ such that

$$\delta(x) = \delta_{\rho^x}(x) \text{ for all } x \geq 0.$$

and $\rho^x \in \mathcal{R}$, where $\mathcal{R}$ is any of the reinsurance families introduced in Definition 2.1. Let us construct the function ρ for any of the families $\mathcal{R}_P$, $\mathcal{R}_{XL}$, and $\mathcal{R}_A$ (the case $\mathcal{R} = \mathcal{R}_F$ is straightforward). Since the optimal survival probability function δ is a viscosity solution of (2.13), analogously to Definition 5.6 (but putting $c = 0$), we can define

$$\hat{\delta}(x) = \inf_{R\in\mathcal{R}} \left\{ \frac{\beta\delta(x) - \beta\mathcal{I}_R(\delta)(x)}{p_R} \right\}.$$

As in Lemma 5.6, we obtain that $\hat{\delta}$ is well defined, nonnegative, Borel measurable, and

$$\sup_{R\in\mathcal{R}}\{p_R\hat{\delta}(x) - \beta\delta(x) + \beta\mathcal{I}_R(\delta)(x)\} = 0,$$

for any $R \in \mathcal{R}$ and $x \geq 0$. Moreover, $\hat{\delta} = \delta'$ at all the points where δ is differentiable (i.e., a.e. in $\mathbf{R}_+$).

As in Proposition 5.6, we obtain that the optimal stationary reinsurance policy in the family of all the proportional retained loss functions satisfies $R^*_x(\alpha) = b^*(x)\alpha$ where

$$b^*(x) = \max(\arg\max_{b\in[0,1]} g_P(\delta, x, b)), \tag{5.24}$$

in the family of all the excess-of-loss retained loss functions satisfies $R^*_x(\alpha) = a^*(x) \wedge \alpha$, where

$$a^*(x) = \max(\arg \max_{a\in[0,x]\cup\{\infty\}} g_{XL}(\delta, x, a)) \tag{5.25}$$

and in the family of all the retained loss functions is given by

$$R^*_x(\alpha) = \max\left(\arg\max_{s\in[0,\alpha]}\left((1+\eta_1)\hat{\delta}(x)s + \delta(x-s)\right)\right). \tag{5.26}$$

Let us show that $p_{R_x^*}$ is nonnegative in all the cases: since δ is increasing, we have

$$p_{R_x^*}\hat{\delta}(x) = \beta\delta(x) - \beta\mathcal{I}_R(\delta)(x) \geq 0,$$

and so we obtain $p_{R_x^*} \geq 0$ because $\hat{\delta} \geq 0$. Note that it is not obvious that $p_{R_x^*} > 0$.

We define $\rho^x(\alpha) = R_x^*(\alpha)$; as in Proposition 5.8, ρ is Borel measurable, but we have to test in each example whether $p_{\rho^x} > 0$ for all $x \geq 0$. If the claim-size distribution F is continuous, Schmidli [57] proved in Lemma 2.10 that this property holds for the families $\mathcal{R}_P$ and $\mathcal{R}_{XL}$.

Let us study now the optimal retain loss functions (5.26) and (5.18) in the family $\mathcal{R}_A$; we focus ourselves in the dividend problem (5.26), the same results hold for the survival probability problem (5.18) replacing δ by V.

Consider the function $g(s) = (1 + \eta_1)\hat{\delta}(x)s + \delta(x - s)$; the optimal retain loss function R_x^* will satisfy

$$R_x^*(\alpha) = \alpha \tag{5.27}$$

if, and only if, the maximum of the function g in the interval $[0, \alpha]$ is reached in the upper limit α. In order to characterize this point, we can use a version of "the rising sun lemma" (see, for instance, Sect. 1.6 of [61]). Imagine that the graph of g is illuminated by the sun shining horizontally from the left, then the points α where (5.27) holds are the one illuminated by the sun, and the points in the shadow form an open set O which could be written as the union of at most countably many disjoint nonempty open intervals, that is $O = \bigcup_n (r_{1,n}, r_{2,n})$. So, we have that

$$R_x^*(\alpha) = \begin{cases} r_{1,n} & \text{if } x \in (r_{1,n}, r_{2,n}) \\ \alpha & \text{if } x \notin O. \end{cases}$$

Moreover, we have the following properties:

- Since $g(x) > g(x^-)$ and $g(s) = (1 + \eta_1)\hat{\delta}(x)s$ for $s > x$, we have that O is bounded and there exist $\alpha_2 > 0$ such that $(x, x + \alpha_2) \subset O$ (therefore, O is nonempty). So the optimal reinsurance policy always involves some reinsurance, but it is better not to buy reinsurance protection for large claims.
- If $x = 0$, the optimal retain loss function is

$$R_0^*(\alpha) = \begin{cases} 0 \text{ if } \alpha < \delta(0)/((1 + \eta_1)\hat{\delta}(0)) \\ \alpha \text{ if } \alpha \geq \delta(0)/((1 + \eta_1)\hat{\delta}(0)). \end{cases}$$

- If V is differentiable at x, we have that $\hat{\delta}(x) = \delta'(x)$ and then

$$(1 + \eta_1)\delta'(x)s + \delta(x - s) > \delta(x)$$

  for s small enough. We conclude that there exists $\alpha_1 > 0$ such that $R_x^*(\alpha) = \alpha$ for $\alpha \leq \alpha_1$. So, it is better not to buy reinsurance protection for small claims.

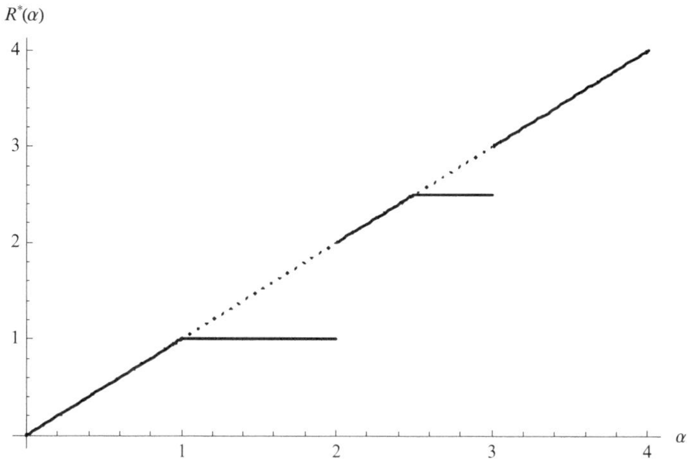

Fig. 5.2 Optimal retain loss function

In Fig. 5.2, we show the graph of the retain loss function corresponding to the open set $O = (1,2) \cup (2.5,3)$.

5.6 Optimal Investment Control for Survival Probability

Here we address the existence of an optimal stationary investment control g in the problem of maximizing the survival probability with investment defined in (2.28). That is the existence of a function $g : \mathbf{R}_+ \to \Gamma$ such that $g(x)x$ is Lipschitz and

$$\delta(x) = \delta^{\overline{\gamma}^g}(x) \text{ for all } x \geq 0,$$

where $\overline{\gamma}^g \in \Pi_x^\gamma$ was introduced in Remark 2.3.

From Theorem 5.5 in [10], we have that δ is twice continuously differentiable, that the optimal stationary investment control is given by

$$g^*(x) = \min\{\hat{\gamma}, 2\frac{\beta\delta(x) - \beta\mathcal{I}(\delta)(x) - p\delta'(x)}{r\hat{\gamma}x\delta'(x)}\} \text{ for } x > 0 \tag{5.28}$$

and that $g^*(0) = \hat{\gamma}$ provided that the function $g^*(x)x$ is Lipschitz in $\mathbf{R}_+$. Moreover, $g^*(x) = \hat{\gamma}$ for small values of x.

5.7 Comments and References

Let us first mention some references on optimal strategies for the classical risk model. For the optimal survival probability problem with reinsurance, see [34, 55, 57]. For the one with investments, see [10, 32]. In 1969, Gerber [29] saw that band strategies are optimal in the bare dividend problem for any claim-size distribution and in [9] it is proved that this result also holds in the reinsurance case. In the problem of dividend payments with investments, [11] showed that limit band strategies are optimal.

In [29] it is proved that the optimal dividend strategy in the classical risk model is barrier for exponential distributed claims. Shreve et al. [58] showed that this also holds for the limit diffusion setting. In these cases, the optimal value functions are smooth.

In the setting that the surplus is modeled by a spectrally negative Lévy process, Avram et al. [7] investigated when the optimal dividend policy is a barrier strategy, Loeffen [43] proved that the optimal dividend strategy is barrier in the case of complete monotone jump density and Kyprianou et al. [41] relaxed this condition on the density of the claim-size distribution to log-convexity. Under these conditions, the optimal value function is differentiable and can be written in terms of the scale function of the underlying Lévy process.

Note that in the problem of maximizing dividends with investments, the controlled process under a band investment strategy does not reflect on the points of the set $\mathcal{A}$ in the sense of a Lévy process. For instance, the value function corresponding to a dividend barrier strategy with positive barrier cannot be computed using the techniques of reflected Lévy processes because the diffusion coefficient of the controlled surplus process (2.22) is not constant and depends on the current surplus. So, we have to introduce the rather intricate notion of limit band strategies.

Chapter 6
Numerical Examples

In this chapter we show some examples of the optimal value functions and the optimal strategies for the classical risk model. In these examples, the optimal band strategies have one (barrier) or two bands; we have not found examples with more bands in the unbounded dividend payment case. However, when imposing a ceiling on the rate of dividends, band strategies with infinitely many bands can be found (even with claim-size distributions with bounded density); see [12].

In the simplest case of no reinsurance and no investment, the survival probability function and the optimal dividend function can be obtained from the almost-everywhere solutions of the equations $\mathcal{L}_0 = 0$ and $\tilde{\mathcal{L}}_0 = 0$ defined in (1.13) and (1.22), respectively. In the case that the claim-size distribution function F is a solution of a linear ODE with constant coefficients, the solutions of the equations $\mathcal{L}_0 = 0$ and $\tilde{\mathcal{L}}_0 = 0$ are also solutions of linear ODEs with constant coefficients, and so they have closed form. Let us consider, for instance, the exponential distribution $F_1(x) = 1 - e^{-\lambda x}$ and the gamma distribution $F_2(x) = 1 - (1 + \lambda x)e^{-\lambda x}$. The solutions of $\tilde{\mathcal{L}}_0 = 0$ are also solutions of the ODE

$$pW''(x) - (c + \beta - \lambda p)W'(x) - \lambda c W(x) = 0$$

for the claim-size distribution F_1 and solutions of the ODE

$$pW'''(x) + (2p - c - \beta)W''(x) + (p - 2c - 2\beta)W'(x) - cW(x) = 0$$

for the claim-size distribution F_2; the solutions of $\mathcal{L}_0 = 0$ satisfy the corresponding ODE with $c = 0$. On the other hand, if the claim size is constant, the solutions of $\mathcal{L}_0 = 0$ and $\tilde{\mathcal{L}}_0 = 0$ have a closed formula as well. For example, if the claim size have constant size one, the almost-everywhere solutions of $\mathcal{L}_0 = 0$ are multiples of the function given in (3.3). For general claim-size distributions, these solutions can be obtained numerically.

In the cases with reinsurance and investment control, the almost-everywhere solution of the equations $\sup_{R \in \mathcal{R}} \mathcal{L}_R = 0$, $\sup_{R \in \mathcal{R}} \tilde{\mathcal{L}}_R = 0$, and the classical

P. Azcue and N. Muler, *Stochastic Optimization in Insurance: A Dynamic Programming Approach*, SpringerBriefs in Quantitative Finance, DOI 10.1007/978-1-4939-0995-7_6,

solutions of $\sup_{\gamma \in \Gamma} \mathcal{L}_\gamma = 0$ and $\sup_{\gamma \in \Gamma} \tilde{\mathcal{L}}_\gamma = 0$ (where the operators are defined in (2.7), (2.20), (2.29), and (2.43), respectively) could be obtained numerically by a finite difference scheme.

6.1 Survival Probability

In order to construct the examples of this section, we first obtain numerically solutions of $\sup_{R \in \mathcal{R}} \mathcal{L}_R(W) = 0$ for the reinsurance case and $\sup_{\gamma \in \Gamma} \mathcal{L}_\gamma(W) = 0$ in $\mathbf{R}_+$ for the investment case together with boundary condition $W(0) = 1$. As a second step we obtain $\delta(x)$ as $W(x)/W(\infty)$ and also approximate the optimal controls.

6.1.1 Examples with Reinsurance

In the first example we compare the survival probability function without reinsurance and the optimal survival probability function introduced in Sect. 2.1.1 among the reinsurance families $\mathcal{R}_{XL}, \mathcal{R}_P$, and $\mathcal{R}_A$ with exponential claim-size distribution $F(x) = 1 - e^{-\lambda x}$.We consider the parameters $\lambda = 3$, $\beta = 10$, $\eta = 0.3$, and $\eta_1 = 0.35$. In Fig. 6.1a, we show from bottom to top the survival probability function without reinsurance and the optimal survival probability functions corresponding to the families $\mathcal{R}_P$, $\mathcal{R}_{XL}$, and $\mathcal{R}_A$ respectively. In this example, the optimal stationary reinsurance control ρ in the family $\mathcal{R}_A$ has the form

$$\rho^x(\alpha) = \begin{cases} \alpha & \text{if } \alpha \in [0, r_1^*(x)] \cup [r_2^*(x), \infty) \\ r_1^*(x) & \text{if } \alpha \in (r_1^*(x), r_2^*(x)), \end{cases} \tag{6.1}$$

where $0 < r_1^*(x) < r_2^*(x)$ are called the optimal levels. We show the optimal retained proportion $b^*(x)$ defined in (5.24) for the family $\mathcal{R}_P$, the optimal retention level $a^*(x)$ defined in (5.25) for the family $\mathcal{R}_{XL}$, and the optimal levels $r_1^*(x)$ and $r_2^*(x)$ corresponding to the family $\mathcal{R}_A$ in Figs. 6.1b, c, and d, respectively. The optimal retention level a^* is infinite for small surpluses (so it is not shown in the graph), and the dotted line is the identity function.

In the second example, we compare the optimal survival probability functions with and without reinsurance in the case where the claims have constant size one. The value function with reinsurance is obtained numerically by a finite difference scheme. Since the only choice that matters is the part of the claim that the insurance company pays when the size of the claim is $\alpha = 1$, the problems with reinsurance families $\mathcal{R}_{XL}, \mathcal{R}_P$, and $\mathcal{R}_A$ coincide. We consider the parameters $\beta = 10$, $\eta = 0.3$, and $\eta_1 = 0.35$. In Fig. 6.2a, we show the survival probability function without reinsurance on the bottom and the optimal survival probability function on the top;

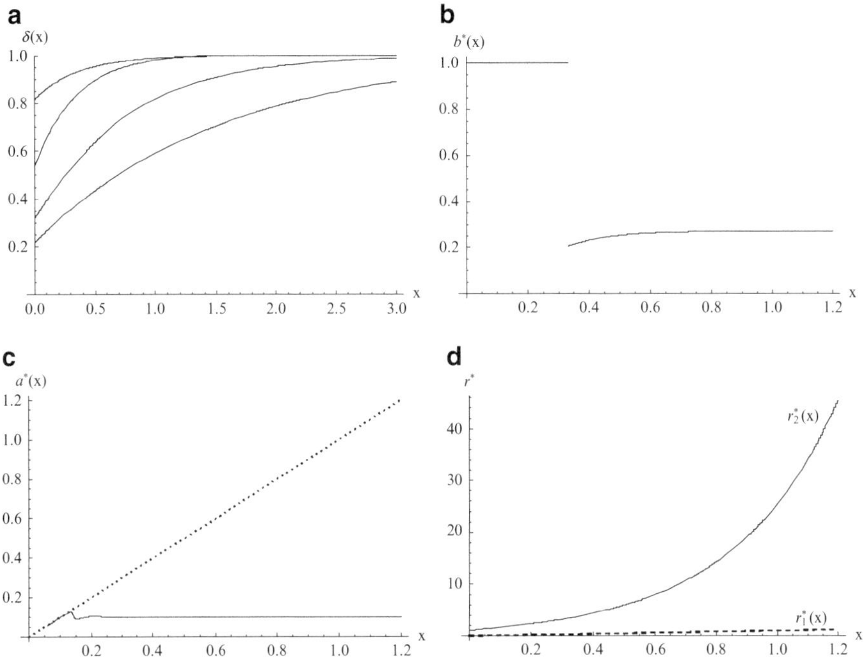

Fig. 6.1 (**a**) Survival probability functions with reinsurance. (**b**) Optimal retained proportion. (**c**) Optimal retention level for excess-of-loss reinsurance. (**d**) Optimal retention levels for general reinsurance

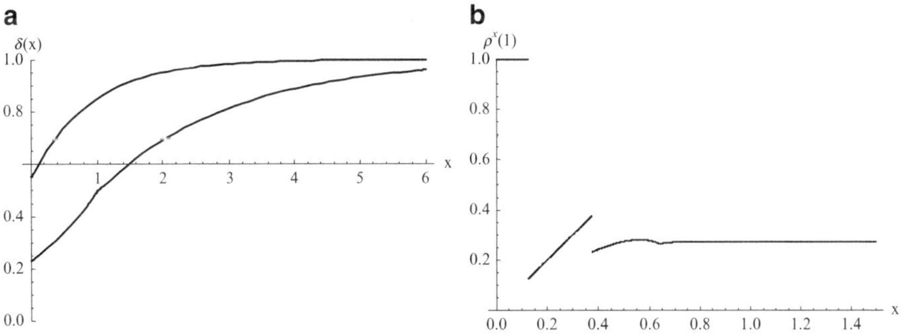

Fig. 6.2 (**a**) Optimal survival probability for claims of size one with and without reinsurance. (**b**) Optimal reinsurance control for claims of size one

note that the optimal survival probability function is continuously differentiable, but the survival probability function without reinsurance is not differentiable at one. In Fig. 6.2b, we show the graph of the function $\rho^x(1)$ where ρ is the optimal stationary reinsurance control; we observe from this graph that the optimal reinsurance control

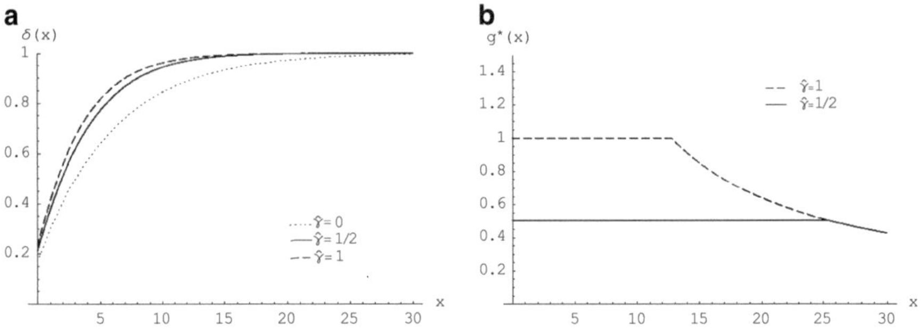

Fig. 6.3 (**a**) Optimal survival probability with investment with exponential distribution. (**b**) Optimal investment controls with exponential distribution

depends on the surplus x in the following way: for $0 \leq x \leq 0.13$, take no reinsurance ($\rho^x(1) = 1$); for $0.13 < x \leq 0.38$ take a reinsurance contract in such a way that the remaining surplus after a claim payment is zero ($\rho^x(1) = x$); and finally for $x > 0.38$, the proportion of the claim paid by the insurer goes asymptotically to 0.29.

6.1.2 Examples with Investments

We consider in the first example of this section, the exponential claim-size distribution $F(x) = 1 - e^{-x}$. We compare the survival probability function without investment with the optimal survival probability functions corresponding to $\hat{\gamma} = 1$ (gearing is not allowed) and $\hat{\gamma} = 1/2$ (at most half of the surplus can be invested in the risky asset) introduced in Sect. 2.2.1. We take the parameters $\beta = 1$, $r = 0.04$, $\sigma = 0.1$, and $p = 1.2$. The optimal survival probability functions are obtained numerically. These results are shown in Fig. 6.3a. In Fig. 6.3b, we show the graphs of the optimal investment stationary controls; the constraints $\hat{\gamma}$ are binding for small surpluses but not for large ones; in fact g^* goes to zero at infinity (see Example 6.1 in [10] for details about the asymptotic behavior). It can be seen in these examples that the functions $xg^*(x)$ are indeed Lipschitz (with Lipschitz constant that coincides with $\hat{\gamma}$).

In the second example, we consider Pareto claim-size distribution $F(x) = 1 - 1/(1 + x)^2$ and the same parameters of the previous example. We compare the survival probability function without investment with the optimal survival probability functions corresponding to $\hat{\gamma} = 1$ and $\hat{\gamma} = 2$. We show the graphs of the optimal survival probability functions in Fig. 6.4a and the graph optimal investment stationary controls in Fig. 6.4b. In the case $\hat{\gamma} = 1$, the constraint is always binding; this means that the optimal investment strategy consists in investing all the surplus in the risky assets. In the case $\hat{\gamma} = 2$, the constraint is binding only for small surpluses but not for large ones; indeed, it can be seen that g^* goes to $4/3$ at infinity (see Example 6.2 in [10] for details about the asymptotic behavior).

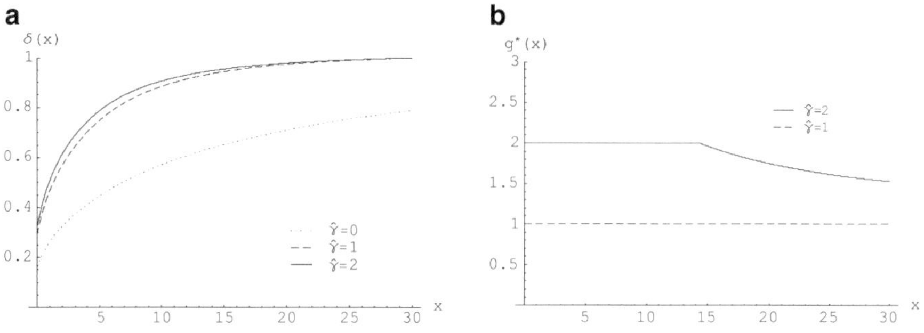

Fig. 6.4 (**a**) Optimal survival probability with investment with Pareto distribution. (**b**) Optimal investment controls with Pareto distribution

6.2 Optimal Dividends

For the problems of optimal dividend payments, we need a procedure to calculate the optimal value functions and the optimal band partitions $\mathcal{P}^* = (\mathcal{A}^*, \mathcal{B}^*, \mathcal{C}^*)$ defined in Chap. 5. We assume that $\mathcal{A}^*$ is finite (and so $\mathcal{B}^*$ and $\mathcal{C}^*$ have finitely many connected components); let us call $n_0 = \#\mathcal{A}^*$. If $\mathcal{A}^*$ were infinite, the same algorithm would provide a way to approximate the optimal value function in the case that the points of $\mathcal{A}^*$ do not accumulate (i.e., $\mathcal{A}^* \cap [0, m]$ is finite for any $m \in \mathbf{N}$).

We say that a band strategy is an n-band strategy if $\#\mathcal{A} = n$. For instance, a 1-band strategy is a barrier strategy. An n-band partition $\mathcal{P}_n = (\mathcal{A}_n, \mathcal{B}_n, \mathcal{C}_n)$ consists in sets $\mathcal{A}_n = \{a_1, .., a_n\}$, $\mathcal{B}_n = \bigcup_{i=1}^{n-1}(a_i, b_i] \cup (a_n, \infty)$, and $\mathcal{C}_n = [0, a_1) \cup \bigcup_{i=1}^{n-1}(b_i, a_{i+1})$ where $0 \leq a_1 < b_1 < a_2 < b_2 < \cdots < a_n$. The main idea of this procedure is to look first for the best 1-band strategy and then to see whether its value function is a viscosity solution of the corresponding HJB equation. If this is the case, by the verification theorem, we have obtained the optimal band strategy. Otherwise, we look for the best 2-band strategy, and so on until we reach n_0. We construct iteratively the best n-band strategy from the best $(n-1)$-band strategy by solving an optimization problem with 2 variables.

For simplicity we address first the simplest case without reinsurance or investment. By Proposition 5.2, the value function $V_{\mathcal{P}_n}$ can be obtained as follows:

- In $[0, a_1]$, $V_{\mathcal{P}_n}$ is the unique almost-everywhere solution of $\tilde{\mathcal{L}}_0 = 0$ with boundary condition $\Lambda(V_{\mathcal{P}_n})(a_1) = 0$.
- In (b_i, a_{i+1}), $V_{\mathcal{P}_n}$ is the unique almost-everywhere solution of $\tilde{\mathcal{L}}_0 = 0$ with boundary condition $\Lambda(V_{\mathcal{P}_n})(a_{i+1}) = 0$.
- In $(a_i, b_i]$, $V_{\mathcal{P}_n}(x) = V_{\mathcal{P}_n}(a_i) + x - a_i$.
- In (a_n, ∞), $V_{\mathcal{P}_n}(x) = V_{\mathcal{P}_n}(a_n) + x - a_n$.

Let $\mathcal{P}_n^* = (\mathcal{A}_n^*, \mathcal{B}_n^*, \mathcal{C}_n^*)$ be the best n-band partition for $1 \leq n \leq n_0$. By Proposition 5.3, the value function $V_{\mathcal{P}_n^*}$ coincides with the optimal value function V

up to the n-th point of $\mathcal{A}^*$. Therefore, if $\mathcal{A}^* = \{a_1^*, \ldots, a_{n_0}^*\}$, $\mathcal{B}^* = \bigcup_{i=1}^{n_0-1}(a_i^*, b_i^*] \cup (a_{n_0}^*, \infty)$, and $\mathcal{C}^* = [0, a_1^*) \cup \bigcup_{i=1}^{n_0-1}(b_i^*, a_{i+1}^*)$, then $\mathcal{A}_n^* = \{a_1^*, \ldots, a_n^*\}$, $\mathcal{B}_n^* = \bigcup_{i=1}^{n-1}(a_i^*, b_i^*] \cup (a_n^*, \infty)$ and $\mathcal{C}_n^* = [0, a_1^*) \cup \bigcup_{i=1}^{n-1}(b_i^*, a_{i+1}^*)$. This property allows us to obtain iteratively the best n-band strategy.

Let us describe the algorithm. The value function $V_{\mathcal{P}_1}$ at zero depends only on the barrier level a_1, so we can find the best 1-band partition (barrier) $\mathcal{P}_1^*$ looking for the value a_1 which maximizes $V_{\mathcal{P}_1}(0)$; this value turns out to be a_1^*. We check whether $V_{\mathcal{P}_1^*}$ is a viscosity solution of the HJB equation; if this is the case, we are done, and if it is not, we look for the best 2-band partition $\mathcal{P}_2^*$. We find $\mathcal{P}_2^*$ in the following way: given any pair of values $b_1 < a_2$ with $b_1 > a_1^*$ we consider the value function $V_{\mathcal{P}_2^{b_1,a_2}}$ corresponding to the 2-band partition $\mathcal{P}_2$ with $\mathcal{A}_2 = \{a_1^*, a_2\}$ and $\mathcal{B}_2 = \left(a_1^*, b_1\right] \cup (a_2, \infty)$. We look for the maximum of $v(b_1, a_2) = V_{\mathcal{P}_2^{b_1,a_2}}(a_2) - a_2$. By Proposition 5.3, we have that this maximum is attained at (b_1^*, a_2^*). Again, we check whether the value function of $\mathcal{P}_2^*$ is a viscosity solution of the HJB equation; if this is the case, we are done, and if it is not, we look for $\mathcal{P}_3^*$ (which gives the values b_2^* and a_3^*) and so on.

In the problem with reinsurance introduced in (2.16), the procedure is similar. Given a finite band partition $\mathcal{P}_n$, we construct numerically a function $\tilde{W}_{\mathcal{P}_n}$ satisfying

- In $[0, a_1]$, $\tilde{W}_{\mathcal{P}_n}$ is the unique almost-everywhere solution of $\sup_{R \in \mathcal{R}} \tilde{\mathcal{L}}_R = 0$ with boundary condition $\sup_{R \in \mathcal{R}} \Lambda_R(\tilde{W}_{\mathcal{P}_n})(a_1) = 0$.
- In (b_i, a_{i+1}), $\tilde{W}_{\mathcal{P}_n}$ is the unique almost-everywhere solution of $\sup_{R \in \mathcal{R}} \tilde{\mathcal{L}}_R = 0$ with boundary condition $\sup_{R \in \mathcal{R}} \Lambda_R(\tilde{W}_{\mathcal{P}_n})(a_{i+1}) = 0$.
- In $(a_i, b_i]$, $\tilde{W}_{\mathcal{P}_n}(x) = \tilde{W}_{\mathcal{P}_n}(a_i) + x - a_i$.
- In (a_n, ∞), $\tilde{W}_{\mathcal{P}_n}(x) = \tilde{W}_{\mathcal{P}_n}(a_n) + x - a_n$.

The function $\tilde{W}_{\mathcal{P}_n}$ can be regarded as the value function of the n-band strategy corresponding to the band partition $\mathcal{P}_n$ and the best possible reinsurance policy. We can find the best n-band partition $\mathcal{P}_n^*$ for $n \le n_0$. By Remark 5.5, we have that $\tilde{W}_{\mathcal{P}_n^*} = V$ in $[0, a_n]$ and so $\tilde{W}_{\mathcal{P}_n^*}$ is the value function of the n- band strategy corresponding to the n-band partition $\mathcal{P}_n^*$ and stationary reinsurance control

$$\rho_n^x = \begin{cases} \arg\max_{R \in \mathcal{R}} \Lambda_R(V)(x) \text{ if } x \in \mathcal{A}_n^* \\ \arg\max_{R \in \mathcal{R}} \tilde{\mathcal{L}}_R(V)(x) \text{ if } x \in \mathcal{C}_n^*. \end{cases}$$

Note that ρ_n^x coincide with the optimal stationary reinsurance control ρ^x in $\mathcal{A}_n^* \cup \mathcal{C}_n^*$.

In the case with investments introduced in (2.33), the procedure is analogous to the former case. Given a finite band partition $\mathcal{P}_n$, we construct numerically a continuous function $\overline{W}_{\mathcal{P}_n}$ satisfying

- $\sup_{\gamma \in \Gamma} \tilde{\mathcal{L}}_\gamma = 0$ in $[0, a_1]$ with boundary condition $\overline{W}'_{\mathcal{P}_n}(a_1) = 1$,
- $\sup_{\gamma \in \Gamma} \tilde{\mathcal{L}}_\gamma = 0$ in (b_i, a_{i+1}) with boundary conditions $\overline{W}_{\mathcal{P}_n}(b_i^+) = \overline{W}_{\mathcal{P}_n}(b_i)$ and $\overline{W}'_{\mathcal{P}_n}(a_{i+1}) = 1$,

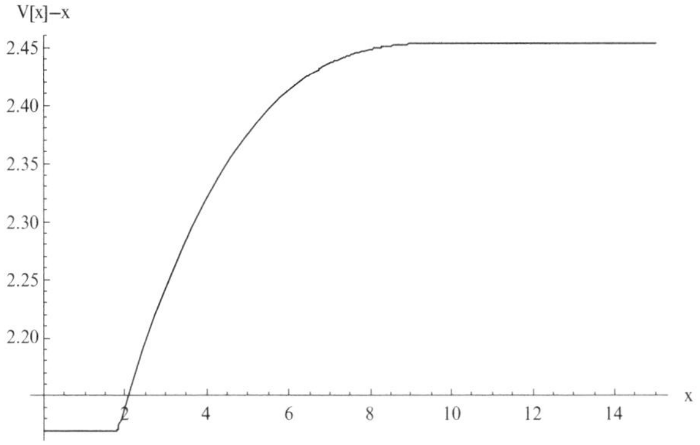

Fig. 6.5 Optimal value function with gamma distribution

- $\overline{W}_{\mathcal{P}_n}(x) = \overline{W}_{\mathcal{P}_n}(a_i) + x - a_i$ in $(a_i, b_i]$,
- $\overline{W}_{\mathcal{P}_n}(x) = \overline{W}_{\mathcal{P}_n}(a_n) + x - a_n$ in (a_n, ∞).

The function $\overline{W}_{\mathcal{P}_n}$ can be regarded as the value function of the limit n-band dividend strategy corresponding to the band partition $\mathcal{P}_n$ and the best possible investment policy. As before, we can find the best limit n-band partition $\mathcal{P}_n^*$ for $n \leq n_0$. By Remark 5.8, we have that $\overline{W}_{\mathcal{P}_n^*} = V$ in $[0, a_n]$ and so $\overline{W}_{\mathcal{P}_n^*}$ is the value function of the limit n- band strategy corresponding to the n-band partition $\mathcal{P}_n^*$ and stationary investment control $g_n^*(x) = \arg\max_{\gamma \in \Gamma} \tilde{\mathcal{L}}_\gamma(V)(x)$ in $\mathcal{C}_n^*$. Note that g_n^* coincide with the optimal stationary investment control g^* in $\mathcal{C}_n^*$.

6.2.1 Dividends (Bare Case)

We first consider the dividend payments problem introduced in Sect. 1.2 with claim-size distribution gamma $F(x) = 1 - (1 + x)e^{-x}$ and parameters $\beta = 10$, $c = 0.1$, $\eta = 0.07$ (and so $p = 21.4$). The optimal band partition is 2-band with $\mathcal{A}^* = \{0, 10.22\}$, $\mathcal{B}^* = (0, 1.083] \cup (10.22, \infty)$, and $\mathcal{C}^* = (1.083, 10.22)$. In Fig. 6.5 we show the graph of $V(x) - x$ where V is the optimal value function; this function is not differentiable at $x = 1.083$ and it is not concave.

In Fig. 6.6, we show the structure of the optimal band partitions for this gamma claim-size distribution depending on the parameters η and c/β. In the region on the left (dark grey), the optimal strategy is barrier with positive level $a > 0$, and in the region on the right (light grey) the optimal strategy is barrier with $a = 0$,; that is the optimal strategy is to pay out immediately all the surplus and then to pay the incoming premium p as dividends up to the arrival time of the first claim (i.e., the ruin time). In the intermediate region (white), the optimal strategy has two or more bands; the previous example falls in this region.

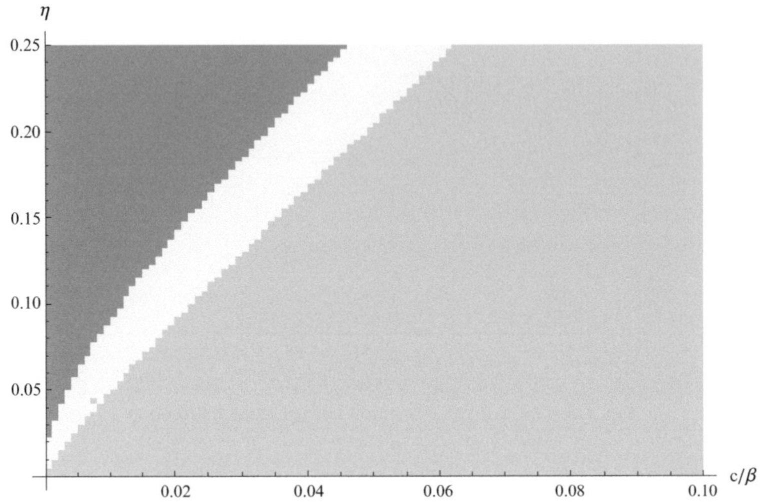

Fig. 6.6 Structure of optimal band partitions

6.2.2 Dividends with Reinsurance

We consider in this section the problem introduced in Sect. 2.1.2. In the first example, we take the exponential claim-size distribution $F(x) = 1 - e^{-3x}$ and parameters $c = 0.2$, $\beta = 10$, $\eta = 0.3$, and $\eta_1 = 0.35$. We first compare the optimal value functions without reinsurance and with reinsurance in the families $\mathcal{R}_{XL}$, $\mathcal{R}_P$, and $\mathcal{R}_A$. In all the cases, the optimal value functions are differentiable and the optimal band strategies are barrier: $\mathcal{A}^* = \{2.74\}$ in the case without reinsurance and with reinsurance in the family $\mathcal{R}_P$, $\mathcal{A}^* = \{2.48\}$ in the family $\mathcal{R}_{XL}$ and $\mathcal{A}^* = \{2.14\}$ in the family $\mathcal{R}_A$. In Fig. 6.7, we show from bottom to top the graphs of $V(x) - x$ for the cases without reinsurance and the families $\mathcal{R}_P$, $\mathcal{R}_{XL}$, and $\mathcal{R}_A$ respectively; the optimal retained proportion defined in (5.14) for the family $\mathcal{R}_P$ is $b^* = 1$ for all surpluses $x \geq 0$ and so the optimal value function in the family $\mathcal{R}_P$ coincides with the one without reinsurance.

As in Sect. 6.1.1, the optimal stationary reinsurance control in the family $\mathcal{R}_A$ has the form (6.1). In Fig. 6.8a, we show the optimal retention level $a^*(x)$ defined in (5.16) for the family $\mathcal{R}_{XL}$ and in Fig. 6.8b, the optimal levels $r_1^*(x)$ and $r_2^*(x)$ as defined in (6.1) corresponding to the family $\mathcal{R}_A$. As in the first example of Sect. 6.1.1, the optimal retention level a^* is infinite for small surpluses and the dotted line is the identity function. In each family, we show the optimal reinsurance controls in $\mathcal{C}^* \cup \mathcal{A}^*$, which are the relevant values of the surplus.

In the second example, we compare the optimal value functions with and without reinsurance in the case where the claims have constant size one. We consider the parameters $\beta = 10$, $c = 0.2$, $\eta = 0.3$, and $\eta_1 = 0.35$. In Fig. 6.9a, we show

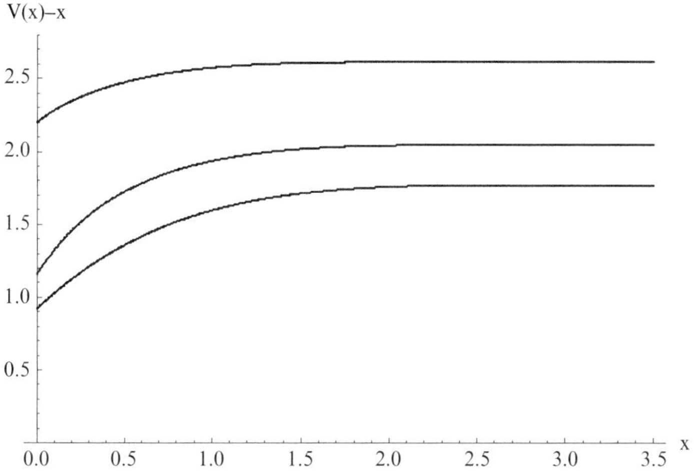

Fig. 6.7 Optimal value functions with reinsurance and exponential distribution

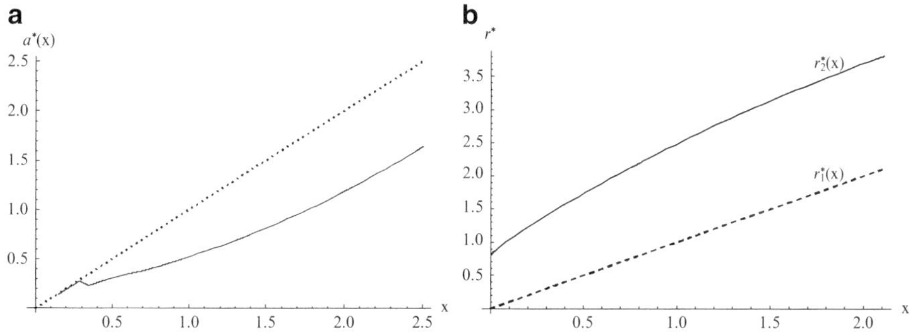

Fig. 6.8 (**a**) Optimal retention level for excess-of-loss reinsurance with exponential distribution. (**b**) Optimal retention levels for general reinsurance with exponential distribution

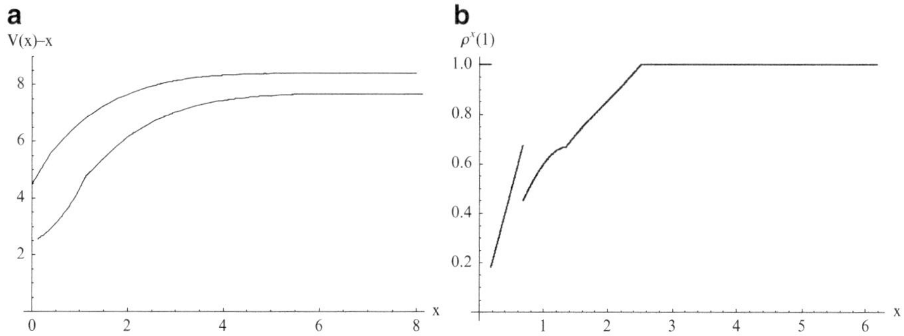

Fig. 6.9 (**a**) Optimal value functions with and without reinsurance and claims of size one. (**b**) Optimal reinsurance control for claims of size one

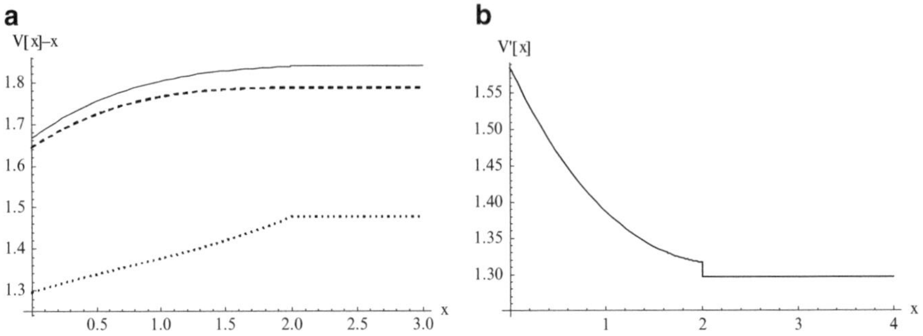

Fig. 6.10 (**a**) Optimal value functions for the problem with two retained loss functions. (**b**) Derivative of the optimal value function in the third family

the graphs of $V(x) - x$ for the optimal value functions V: *at the top* the case with reinsurance, and *at the bottom* the one without reinsurance. Note that in the case without reinsurance, the optimal value function is not differentiable at $x = 1$. The optimal band strategies are barrier. In Fig. 6.9b, we show the graph of the function $\rho^x(1)$ where ρ is the optimal stationary reinsurance control; we observe from this graph that the optimal stationary reinsurance control depends on the surplus x in the following way: for small and large surpluses $x \in [0, 0.18) \cup [2.51, \infty)$, take no reinsurance ($\rho^x(1) = 1$); for $x \in [0.18, 0.67]$, take a reinsurance contract in such a way that the remaining surplus after a claim payment is zero ($\rho^x(1) = x$); and finally for $x \in [0.67, 2.51)$, the proportion of the claim paid by the insurance company is smaller than one.

We define for the third example two retained loss functions $R_0(\alpha) = \alpha$ and $R_1(\alpha) = (2 \wedge \alpha)$. We consider three finite families $\mathcal{R}_0 = \{R_0\}$ (which corresponds to the case without reinsurance), $\mathcal{R}_1 = \{R_1\}$, and $\mathcal{R}_2 = \{R_0, R_1\}$. The claim-size distribution is $F(x) = 1 - e^{-x}$ and the parameters are $\beta = 1$, $c = 0.1$, and $\eta = \eta_1 = 0.5$. We show from bottom to top in Fig. 6.10a the graphs of $V(x) - x$ for $\mathcal{R}_1$, $\mathcal{R}_0$, and $\mathcal{R}_2$ respectively. The optimal value functions are not differentiable at $x = 2$ in the families $\mathcal{R}_1$ and $\mathcal{R}_2$ (this is because F_{R_1} is not continuous at $x = 2$). In the three families, the optimal band partitions are barrier with $\mathcal{A}^* = \{2.21\}$ in $\mathcal{R}_0$ and $\mathcal{A}^* = \{2\}$ in both families $\mathcal{R}_1$ and $\mathcal{R}_2$; the optimal stationary reinsurance control for the family $\mathcal{R}_2$ is

$$\rho^x = \begin{cases} R_0 \text{ if } x < 2 \\ R_1 \text{ if } x = 2. \end{cases}$$

We show in Fig. 6.10b the derivative of the optimal value function corresponding to $\mathcal{R}_2$. Since the optimal value functions in the families $\mathcal{R}_1$ and $\mathcal{R}_2$ are not smooth at the threshold point $a_1^* = 2$, the smooth fitting principle does not hold.

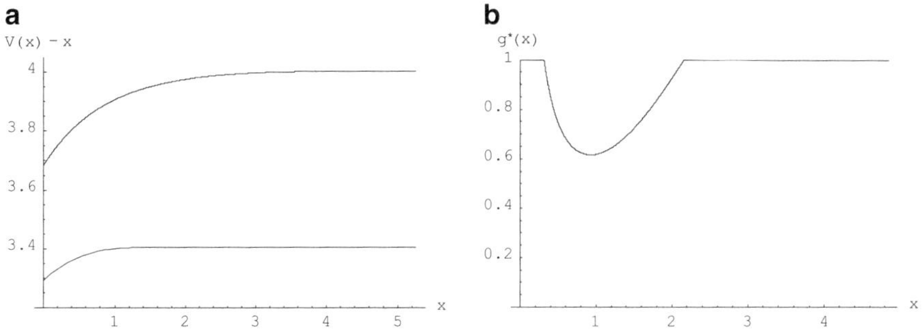

Fig. 6.11 (**a**) Optimal value functions for the problem with investment for exponential distribution. (**b**) Optimal investment control for exponential distribution

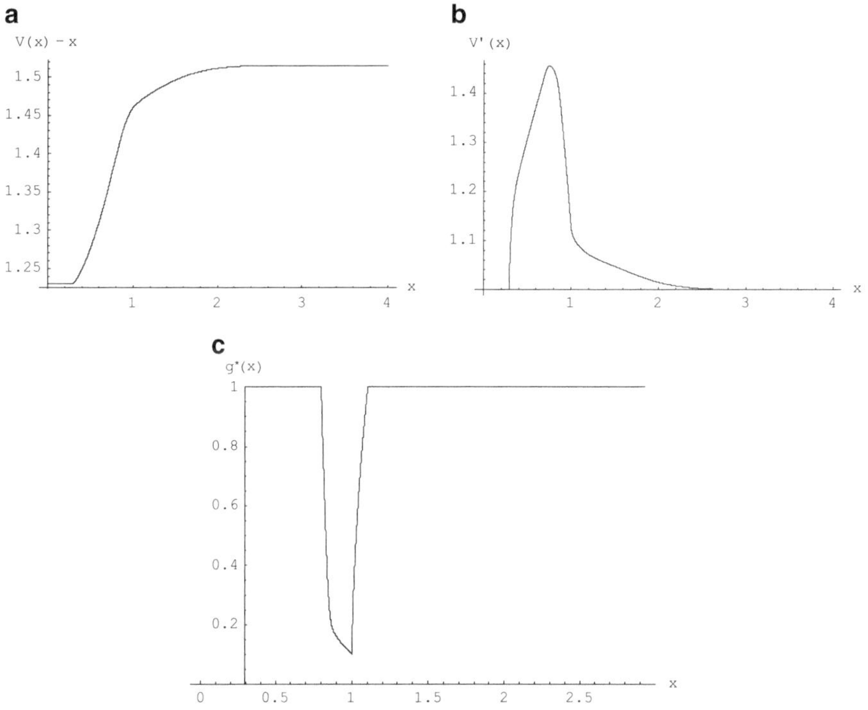

Fig. 6.12 (**a**) Optimal value functions for the problem with investment for the F distribution. (**b**) Derivative of the optimal value function (**c**) Optimal investment control for the F distribution

6.2.3 Dividends with Investments

We consider in this section the problem introduced in Sect. 2.2.2. In the first example, we take the exponential claim-size distribution $F(x) = 1 - e^{-x}$, $\hat{\gamma} = 1$, and the parameters $p = 4$, $\beta = 1$, $c = 0.5$, $r = 0.3$, $\sigma = 2$. We show in Fig. 6.11a the graph of $V(x) - x$ for the optimal dividend payments problem with investment on the top and the one without investment on the bottom. In both cases, the optimal band partitions are barrier, with $\mathcal{A}^* = \{4.85\}$ in the case with investments and $\mathcal{A}^* = \{1.60\}$ in the one without investment. In Fig. 6.11b, we show the optimal stationary investment control g^* in $\mathcal{C}^*$. Note that $g^* = 1$ for small and large surpluses.

Finally, we consider the following claim-size distribution:

$$F(x) = \begin{cases} 0 & \text{if } x \in [0, 7/10] \\ (10/3)(x - 7/10) & \text{if } x \in (7/10, 1] \\ 1 & \text{if } x > 1, \end{cases}$$

$\hat{\gamma} = 1$ and the parameters $p = 1.6$, $\beta = 1$, $c = 0.3$, $r = 0.2$, $\sigma = 1$. We obtain that the optimal limit band partition has two bands with $\mathcal{A}^* = \{0, 2.93\}$, $\mathcal{B}^* = (0, 0.29] \cup (2.93, \infty)$, and $\mathcal{C}^* = (0.29, 2.93)$. We show in Fig. 6.12a the graph of the function $V(x) - x$, in Fig. 6.12b the derivative of V, and in Fig. 6.12c the optimal stationary investment control g^* in $\mathcal{C}^*$. It can be seen in Fig. 6.12b that V is not twice continuously differentiable at $b_1^* = 0.29$, and so the optimal value function is not a classical solution of the corresponding HJB equation (2.42).

Appendix A
Probability Theory and Stochastic Processes

This appendix contains a summary of some basic concepts and results of probability theory and stochastic processes that are used in this brief. For more details and the proofs, see Feller [27] and Varadhan [65, 66].

A.1 Probability Spaces, σ-Algebras, Probability Functions, and Random Variables

Given a set Ω (sample space), we consider a family Σ of subsets of Ω (set of events). The family Σ is called a *σ-algebra* if it satisfies the following properties:

(1) $\varnothing, \Omega \in \Sigma$.
(2) If $A \in \Sigma$, then $A^c \in \Sigma$.
(3) If $A_n \in \Sigma$ for $n \in \mathbf{N}$, then $\bigcup_{n \in \mathbf{N}} A_n \in \Sigma$.

The following result holds.

Proposition A.1. *Given any family $(\Sigma_s)_{s \in S}$ of σ-algebras, the intersection $\bigcap_{s \in S} \Sigma_s$ is also a σ-algebra.*

If Σ is a σ-algebra of Ω, the pair (Ω, Σ) is called a *measurable space.*

A *random variable* on the measurable space (Ω, Σ) is a function $X : \Omega \to \mathbf{R}$ which satisfies that

$$X^{-1}((-\infty, a]) := \{\omega \in \Omega : X(\omega) \leq a\} \in \Sigma \text{ for all } a \in \mathbf{R}.$$

For example, we have the following straightforward result.

Lemma A.1. *Given any set $A \in \Sigma$, the indicator or characteristic function I_A : $\Omega \to \mathbf{R}$ defined as*

P. Azcue and N. Muler, *Stochastic Optimization in Insurance: A Dynamic Programming Approach*, SpringerBriefs in Quantitative Finance, DOI 10.1007/978-1-4939-0995-7,

$$I_A(\omega) = \begin{cases} 1 \text{ if } \omega \in A \\ 0 \text{ if } \omega \notin A \end{cases}$$

is a random variable.

We use Proposition A.1 to introduce the following σ-algebras: Given a family Φ of subsets of Ω, let us define the *σ-algebra generated by* Φ as

$$\Sigma(\Phi) = \bigcap \{\Sigma : \ \Sigma \text{ is } \sigma\text{-algebra and } \Phi \subset \Sigma\},$$

and given a function $f : \Omega \to \mathbf{R}$, let us define the *$\sigma$-algebra generated by* f by

$$\Sigma(f) = \bigcap \{\Sigma : \ \Sigma \text{ is } \sigma\text{-algebra and } f^{-1}((-\infty, a]) \in \Sigma \text{ for } a \in \mathbf{R}\}.$$

Proposition A.2. *Given a function* $f : \Omega \to \mathbf{R}$, *$f$ is a random variable on the measurable space* $(\Omega, \Sigma(f))$. *On the other hand, if f is a random variable on the measurable space* (Ω, Σ), *then* $\Sigma(f) \subset \Sigma$.

A function $P : \Sigma \to [0, 1]$ is called a *probability function* on (Ω, Σ) if it satisfies the following three properties:

(1) $P(A) \geq 0$ for all $A \in \Sigma$.
(2) $P(\varnothing) = 0$ and $P(\Omega) = 1$.
(3) If $(A_n)_{n\in\mathbf{N}}$ is a sequence of pairwise disjoint sets in Σ, then $P(\bigcup_{n\in\mathbf{N}} A_n) = \sum_{n\in\mathbf{N}} P(A_n)$.

In the case that P is a probability function on (Ω, Σ), the triple (Ω, Σ, P) is called a *probability space*. A property on (Ω, Σ, P) is said to hold *almost everywhere* (a.e.) or to hold for *almost all* $\omega \in \Omega$ if the set of elements $\omega \in \Omega$ where this property does not hold has probability P equal to zero.

A.2 Expectation, Conditional Expectation, and Conditional Probability

Given a positive random variable X in a probability space (Ω, Σ, P), let us define the *expectation*

$$E(X) = \textstyle\int X dP := \sup\{\sum_{i=1}^{n} a_i P(A_i) \text{ st. } \sum_{i=1}^{n} a_i I_{A_i} \leq X \text{ with } a_i \in \mathbf{R}_+ \text{ and } A_i \in \Sigma\}.$$

The definition of expectation can be extended to any random variable X as

$$E(X) = E(X\ I_{\{X\geq 0\}}) - E((-X)I_{\{X<0\}}),$$

in the case that both $E(X\ I_{\{X\geq 0\}})$ and $E((-X)I_{\{X<0\}})$ are finite.

Given a random variable X in a probability space (Ω, Σ, P) and a σ-algebra $\overline{\Sigma} \subset \Sigma$, the *conditional expectation* is defined as the unique function $E(X|\overline{\Sigma}) : \Omega \to \mathbf{R}$ which is measurable with respect to $\overline{\Sigma}$ and satisfies

$$E(E(X|\overline{\Sigma})\, I_A) = E(X\; I_A) \text{ for any } A \in \overline{\Sigma}.$$

The existence of this function follows from the Radon–Nikodym theorem.

Given two random variables X_1 and X_2 on (Ω, Σ, P), we define the *conditional expectation* of X_1 given X_2 as

$$E(X_1|X_2) = E(X_1|\Sigma(X_2)).$$

Taking $A \in \Sigma$, we have that $E(I_A|\overline{\Sigma})$ is constant for $\omega \in A$; we define the *conditional probability* $P(A|\overline{\Sigma})$ as the value of this constant.

Proposition A.3. *Assume that* $A, B \in \Sigma$ *and* $P(B) \neq 0$, *then*

$$P(A|I_B) = \frac{P(A \cap B)}{P(B)}.$$

This value is called the *conditional probability of* A *given* B and it is denoted by $P(A|B)$; it is the probability that the event A will occur, when another event B is known to occur or to have occurred.

A.3 Construction of Probability Spaces

Most of the important examples of probability spaces can be constructed using the Caratheodory Extension Method. Let us define first the notions of algebras and algebra-probability functions: Given a set Ω (sample space), a family $\boldsymbol{A}$ of subsets of Ω is called an *algebra* if it satisfy the following properties:

(1) $\emptyset, \Omega \in \boldsymbol{A}$.
(2) If $A \in \boldsymbol{A}$, then $A^c \in \boldsymbol{A}$.
(3) If $A_1, A_2 \in \boldsymbol{A}$, then $A_1 \cup A_1 \in \boldsymbol{A}$.

A function $P_{\boldsymbol{A}} : \boldsymbol{A} \to [0, 1]$ is an *algebra-probability function* if it satisfy the properties

(1) $P_{\boldsymbol{A}}(A) \geq 0$ for all $A \in \boldsymbol{A}$.
(2) $P_{\boldsymbol{A}}(\emptyset) = 0$ and $P_{\boldsymbol{A}}(\Omega) = 1$.
(3) If $A_1, A_2 \in \boldsymbol{A}$ are disjoint sets, then $P(A_1 \cup A_2) = P(A_1) + P(A_2)$.

Then, the following extension result holds.

Theorem A.1. *Given an algebra* $\boldsymbol{A}$ *and an algebra-probability function* $P_{\boldsymbol{A}}$ *in* Ω, *then there exists a unique probability space* $(\Omega, \Sigma(\boldsymbol{A}), P)$ *such that* $P(A) = P_{\boldsymbol{A}}(A)$ *for all* $A \in \boldsymbol{A}$.

This method allows us to construct the product of two probability spaces. Given the probability spaces $(\Omega_1, \Sigma_1, P_1)$ and $(\Omega_2, \Sigma_2, P_2)$ we can consider the algebra $\boldsymbol{A}$ of all the subsets of $\Omega_1 \times \Omega_2$ that can be written as finite disjoint unions $\bigcup_{k=1}^{n} A_{1,k} \times A_{2,k}$ with $A_{1,k} \in \Sigma_1$ and $A_{2,k} \in \Sigma_2$, and the algebra-probability function $P_{\boldsymbol{A}}$ is defined as

$$P_{\boldsymbol{A}}(\textstyle\bigcup_{k=1}^{n} A_{1,k} \times A_{2,k}) = \sum_{k=1}^{n} P_1(A_{1,k}) P_2(A_{2,k}).$$

The unique extension $(\Omega_1 \times \Omega_2, \Sigma(\boldsymbol{A}), P)$ given by the previous theorem is denoted by $(\Omega_1, \Sigma_1, P_1) \times (\Omega_2, \Sigma_2, P_2)$.

A.4 Stochastic Processes and Filtrations

A *stochastic process* is a family $\overline{X} = (X_t)_{t \in J}$ of random variables. In the remaining of the appendix, the index set J will be $\mathbf{R}$, $\mathbf{R}_+$, or an interval. Note that given $\omega \in \Omega$, the *path process* $t \to X_t(\omega)$ is a function from J to $\mathbf{R}$. The stochastic process is said to be *càdlàg* if the path processes $t \to X_t(\omega)$ are right continuous with left limits for almost all ω. Analogously, the stochastic process is said to be *càglàd* if the path processes are left continuous with right limits for almost all ω.

An increasing family of σ-algebras $\mathcal{F} = (\mathcal{F}_t)_{t \in J}$ is called a *filtration*. We say that $(\Omega, \Sigma, \mathcal{F}, P)$ is a *filtered probability* space if $\mathcal{F}_t \subset \Sigma$ for any $t \in J$. $\mathcal{F}_t$ is called the σ-algebra of *events observable until time* t.

Given a filtration $\mathcal{F}$, let us define

$$\mathcal{F}_t^+ = \textstyle\bigcap_{s>t} \mathcal{F}_s \text{ and } \mathcal{F}_t^- = \Sigma\left(\bigcup_{s<t} \mathcal{F}_s\right).$$

A filtration $\mathcal{F}$ is said to be *right continuous* if $\mathcal{F}_t^+ = \mathcal{F}_t$ and it is said to be *left continuous* if $\mathcal{F}_t^- = \mathcal{F}_t$. A stochastic process is said to be *adapted* with respect to $\mathcal{F}$ if

$$X_s^{-1}((-\infty, a]) \in \mathcal{F}_t \text{ for all } a \in \mathbf{R} \text{ and } s \leq t,$$

and it is said to be *predictable* with respect to $\mathcal{F}$ if

$$X_s^{-1}((-\infty, a]) \in \mathcal{F}_{t^-} \text{ for all } a \in \mathbf{R} \text{ and } s \leq t.$$

Given a stochastic process $\overline{X}$, the filtration *generated by* $\overline{X} = (X_t)_{t \in J}$ is defined as $(\Sigma(X_t))_{t \in J}$.

A.5 Stopping Times

An *stopping time* τ in the filtered probability space $(\Omega, \Sigma, \mathcal{F}, P)$ is a function $\tau : \Omega \to J$ such that

$$\{\omega \in \Omega : \tau \leq t\} \in \mathcal{F}_t \text{ for all } t \in J.$$

We have the following properties.

Proposition A.4.

(a) If τ_1 and τ_2 are stopping times with respect to $\mathcal{F}$, so are $\tau_1 \wedge \tau_2$ and $\tau_1 \vee \tau_2$.
(b) If $(X_t)_{t \in J}$ is adapted and τ is a stopping time with respect to $\mathcal{F}$, then $(X_{t \wedge \tau})_{t \in J}$ is also adapted with respect to $\mathcal{F}$.
(b) If $(X_t)_{t \in J}$ is càdlàg and adapted with respect to $\mathcal{F}$ and $A \subset \mathbf{R}$ is an open set, then the exit time $\tau_A = \inf\{t \in J : X_t \notin A\}$ is a stopping time with respect to $\mathcal{F}$.

Given a stopping time τ in the filtered probability space $(\Omega, \Sigma, \mathcal{F}, P)$, let us define

$$\mathcal{F}_\tau = \{A \in \Sigma \text{ st. } A \cap \{\omega \in \Omega : \tau \leq t\} \in \mathcal{F}_t \text{ for all } t \in J\}.$$

$\mathcal{F}_\tau$ is the σ-algebra of *events observable until* τ.

A.6 Martingales

An adapted stochastic process $\overline{X} = (X_t)_{t \in J}$ in the filtered probability space $(\Omega, \Sigma, \mathcal{F}, P)$ is a *martingale* (respectively, *submartingale*, *supermartingale*) if it satisfies the following properties:

(1) The path processes have left and right limits a.e.
(2) X_t is integrable.
(3) For any $s < t$, we have that $E(X_t|\mathcal{F}_s) = X_s$ a.e. (respectively, $E(X_t|\mathcal{F}_s) \geq X_s$, $E(X_t|\mathcal{F}_s) \leq X_s$).

An adapted stochastic process $\overline{X} = (X_t)_{t \in J}$ in the filtered probability space $(\Omega, \Sigma, \mathcal{F}, P)$ is a *local martingale* (respectively, local submartingale, local supermartingale) if there exists a sequence $(\tau_n)_{n \in \mathbf{N}}$ of stopping times such that τ_n is nondecreasing a.e., $\lim_{n \to \infty} \tau_n = \infty$ a.e., and $X_{t \wedge \tau_n}$ is a martingale (respectively, submartingale, supermartingale) for all $n \in \mathbf{N}$.

The next result is called Doob's optional stopping theorem.

Theorem A.2. *If $\overline{X} = (X_t)_{t \in J}$ is a martingale and $\tau_1 \leq \tau_2$ are bounded stopping times in the filtered probability space $(\Omega, \Sigma, \mathcal{F}, P)$, then*

$$E(X_{\tau_2}|\mathcal{F}_{\tau_1}) = X_{\tau_1}.$$

We have the following corollary.

Corollary A.1. *If* $\overline{X} = (X_t)_{t \in J}$ *is a martingale and* τ *is a stopping time in the filtered probability space* $(\Omega, \Sigma, \mathcal{F}, P)$*, then* $(X_{\tau \wedge t})_{t \in J}$ *is also a martingale.*

A.7 Markov Processes

A Markov process can be thought as a stochastic process whose future probabilities are determined by its most recent values. More precisely, an adapted stochastic process $\overline{X} = (X_t)_{t \in J}$ in the filtered probability space $(\Omega, \Sigma, \mathcal{F}, P)$ is called *Markovian* if

$$P(A|\mathcal{F}_s) = P(A|X_s)$$

for all $A \in \Sigma(X_t)$ and $s < t$.

Bibliography

1. Albrecher, H., Thonhauser, S.: Dividend maximization under consideration of the time value of ruin. Insur. Math. Econ. **41**, 163–184 (2007)
2. Albrecher, H., Thonhauser, S.: Optimality results for dividend problems in insurance. Rev. R. Acad. Cien. Serie A. Mat. **103**, 295–320 (2009)
3. Asmussen, S., Albrecher, H.: Ruin Probabilities. Advanced Series on Statistical Science and Applied Probability, vol. 14. World Scientific, Singapore (2010)
4. Asmussen, S., Taksar, M.: Controlled diffusion models for optimal dividend pay-out. Insur. Math. Econ. **20**, 1–15 (1997)
5. Asmussen, S., Højgaard, B., Taksar, M.: Optimal risk control and dividend distribution policies. Example of excess-of-loss reinsurance. Financ. Stoch. **4**, 299–324 (2000)
6. Avanzi, B.: Strategies for dividend distribution: a review. N. Am. Actuar. J. **13**, 217–251 (2009)
7. Avram, F., Palmowski, Z., Pistorius, M.: On the optimal dividend problem for a spectrally negative Lévy process. Ann. Appl. Probab. **17**, 156–180 (2007)
8. Avram, F., Palmowski, Z., Pistorius, M.: On Gerber-Shiu functions and optimal dividend distribution for a Levy risk-process in the presence of a penalty function. Working Paper. arXiv:1110.4965v2 (2013)
9. Azcue, P., Muler, N.: Optimal reinsurance and dividend distribution policies in the Cramer-Lundberg model. Math. Financ. **15**, 261–308 (2005)
10. Azcue, P., Muler, N.: Optimal investment strategy to minimize the ruin probability of an insurance company under borrowing constraints. Insur. Math. Econ. **44**, 26–34 (2009)
11. Azcue, P., Muler, N.: Optimal investment policy and dividend payment strategy in an insurance company. Ann. Appl. Probab. **20**, 1253–1302 (2010)
12. Azcue, P., Muler, N.: Optimal dividend policies for compound Poisson processes: the case of bounded dividend rates. Insur. Math. Econ. **51**, 26–42 (2012)
13. Bardi, M., Capuzzo-Dolcetta, I.: Optimal Control and Viscosity Solutions of Hamilton-Jacobi-Bellman Equations. Birkhäuser, Boston (1997)
14. Bellman, R.: Dynamic programming and a new formalism in the calculus of variations. Proc. Natl. Acad. Sci. **40**, 231–235 (1954)
15. Benth, F.E., Karlsen, K.H., Reikvam, K.: Portfolio optimization in a Lévy market with intertemporal substitution and transaction costs. Stoch. Stoch. Rep. **74**, 517–569 (2002)
16. Borch, K.: The Mathematical Theory of Insurance. Lexington Books, D.C. Heath and Company, London (1974)
17. Borodin, A.N., Salminen, P.: Handbook of Brownian Motion—Facts and Formulae, 2nd edn. Birkhäuser, Basel (2002)
18. Browne, S.: Optimal investment policies for a firm with a random risk process: exponential utility and minimizing the probability of ruin. Math. Oper. Res. **20**, 937–958 (1995)

P. Azcue and N. Muler, *Stochastic Optimization in Insurance: A Dynamic Programming Approach*, SpringerBriefs in Quantitative Finance, DOI 10.1007/978-1-4939-0995-7,

19. Cai, J., Feng, R., Willmot, G.E.: On the expectation of total discounted operating costs up to default and its applications. Adv. Appl. Probab. **41**, 495–522 (2009)
20. Cramer, H.: On the Mathematical Theory of Risk. Skandia Jubilee Volume, Stockholm (1930)
21. Crandall, M.G., Lions, P.L.: Viscosity solutions of Hamilton-Jacobi equations. Trans. Am. Math. Soc. **277**, 1–42 (1983)
22. Crandall, M.G., Ishii, H., Lions, P.L.: User's guide to viscosity solutions of second order partial differential equations. Bull. Am. Math. Soc. (NS) **27**, 1–67 (1992)
23. Dickson, D.C.M., Waters, H.R.: Some optimal dividends problems. ASTIN Bull. **34**, 49–74 (2004)
24. De Finetti, B.: Su Un'impostazione alternativa della teoria collettiva del rischio. Transactions of the 15th International Congress of Actuaries, vol. 2, pp. 433–443. Mallon, New York (1957)
25. Durrett, R.: Essentials of Stochastic Process. Springer, New York (1999)
26. Edalati, A., Hipp, C.: Solving a Hamilton–Jacobi–Bellman equation with constraints. Stoch. Int. J. Probab. Stoch. Process. **85**, 637–651 (1013)
27. Feller, W.: An Introduction to Probability Theory and Its Applications, vol. 2, 2nd edn. Wiley, New York (1971)
28. Fleming, W.H., Soner, H.M.: Controlled Markov Processes and Viscosity Solutions. Springer, New York (1993)
29. Gerber, H.U.: Entscheidungskriterien Für Den Zusammengesetzten Poisson-Prozess. Bull. Assoc. Suisse des Actuar. **2**, 185–228 (1969)
30. Gerber, H.U., Shiu, E.S.W.: On optimal dividend strategies in the compound Poisson model. N. Am. Actuar. J. **10**, 76–93 (2006)
31. Gerber, H.U., Lin, X.S., Yang, H.: A note on the dividends-penalty identity and the optimal dividend barrier. ASTIN Bull. **36**, 489–503 (2006)
32. Hipp, C., Plum, M.: Optimal Investment for insurers. Insur. Math. Econ. **27**, 215–228 (2000)
33. Hipp, C., Taksar, M.: Stochastic control for optimal new business. Insur. Math. Econ. **26**, 185–192 (2000)
34. Hipp, C., Vogt, M.: Optimal dynamical XL reinsurance. ASTIN Bull. **33**, 193–207 (2003)
35. Højgaard, B., Taksar, M.: Optimal dynamic portfolio selection for a corporation with controllable risk and dividend distribution policy. Quant. Financ. **4**, 315–327 (2004)
36. Iglehart, D.L.: Diffusion Approximations in collective risk theory. J. Appl. Probab. **6**, 285–292 (1969)
37. Ishii, H.: Perron's method for Hamilton-Jacobi equations. Duke Math. J. **55**, 369–384 (1987)
38. Jeanblanc-Picqué, M., Shiryaev, A.N.: Optimization of the flow of dividends. Uspekhi Mat. Nauk. **50**, 25–46 (1995)
39. Karatzas, I., Shreve, S.E.: Brownian Motion and Stochastic Calculus. Springer, Berlin (1991)
40. Kulenko, N., Schmidli, H.: Optimal dividend strategies in a Cramér-Lundberg model with capital injections. Insur. Math. Econ. **43**, 270–278 (2008)
41. Kyprianou, A.E., Rivero, V., Song, R.: Convexity and smoothness of scale functions and de Finetti's control problem. J. Theor. Probab. **23**, 547–564 (2010)
42. Lions, P.L.: Optimal control of diffusion processes and Hamilton-Jacobi-Bellman equations. II. Viscosity solutions and uniqueness. Comm. Part. Differ. Equat. **8**, 1229–1276 (1983)
43. Loeffen, R.L.: On optimality of the barrier strategy in de Finetti's dividend problem for spectrally negative Levy processes. Ann. Appl. Probab. **18**, 1669–1680 (2008)
44. Loeffen, R.L.: An optimal dividends problem with transaction costs for spectrally negative Levy processes. Insur. Math. Econ. **45**, 41–48 (2009)
45. Loeffen, R.L., Renaud, J.F.: De Finetti's optimal dividends problem with an affine penalty function at ruin. Insur. Math. Econ. **46**, 98–108 (2010)
46. Lundberg, F.: Über Die Theorie Der Rückversicherung. Transactions of the VIth International Congress of Actuaries, vol. 1, pp. 877–948 (1909)
47. Mnif, M., Sulem, A.: Optimal risk control and dividend policies under excess of loss reinsurance. Stoch. Int. J. Probab. Stoch. Process. **77**, 455–476 (2005)
48. Niemiro, W., Pokarowski, P.: Tail events of some nonhomogeneous Markov chains. Ann. Appl. Probab. **5**, 261–293 (1995)

49. Øksendal, B., Sulem, A.: Applied Stochastic Control of Jump Diffussions. Springer, Berlin (2007)
50. Pham, H.: Continuous-time Stochastic Control and Optimization with Financial Applications. Stochastic Modelling and Applied Probability, vol. 6. Springer, Berlin (2009)
51. Protter, P.: Stochastic Integration and Differential Equations. Springer, Berlin (1992)
52. Royden, H.L.: Real Analysis, 2nd edn. Macmillan, New York (1968)
53. Sayah, A.: Équations d'Hamilton-Jacobi du premier ordre avec termes intégro différentiels. I. Unicité des solutions de viscosité. Comm. Part. Differ. Equat. **16**, 1057–1074 (1991)
54. Sayah, A.: Équations d'Hamilton–Jacobi du premier ordre avec termes intégro différentiels. II. Existence des solutions de viscosité. Comm. Part. Differ. Equat. **16**, 1075–1093 (1991)
55. Schmidli, H.: Optimal proportional reinsurance policies in a dynamic setting. Scand. Actuar. J. **1**, 55–68 (2001)
56. Schmidli, H.: On minimizing the ruin probability by investment and reinsurance. Ann. Appl. Probab. **12**, 890–907 (2002)
57. Schmidli, H.: Stochastic Control in Insurance. Springer, New York (2008)
58. Shreve, S.E., Lehoczky, J.P., Gaver, D.P.: Optimal consumption for general diffusions with absorbing and reflecting barriers. SIAM J. Control Optim. **22**, 55–75 (1984)
59. Soner, H.M.: Optimal control of jump-Markov processes and viscosity solutions. Stochastic Differential Systems, Stochastic Control Theory and Applications. IMA Volumes in Mathematics and Its Applications, vol. 10, pp. 501–511. Springer, New York (1988)
60. Souganidis, P.E., Soner, H.M., Evans, L.C., Bardi, M., Crandall, M.G.: Viscosity Solutions and Applications. Lectures Notes in Mathematics, vol. 1660. Springer, Berlin (1997)
61. Tao, T.: An Introduction to Measure Theory. Graduate Studies in Mathematics. American Mathematical Society, Providence (2011)
62. Teugels, J.L.: Reinsurance actuarial aspects. Report 2003–006. Eurandom, Technical University of Eindhoven, Eindhoven, The Netherlands (2003)
63. Thonhauser, S., Albrecher, H.: Optimal dividend strategies for a compound Poisson risk process under transaction costs and power utility. Stoch. Models **27**, 120–140 (2011)
64. Touzi, N.: Optimal Stochastic Control, Stochastic Target Problems, and Backward SDE. Fields Institute Monographs, vol. 29. Springer, New York (2013)
65. Varadhan, S.R.S.: Probability Theory. Courant Lecture Notes, vol. 7. AMS, Providence (2001)
66. Varadhan, S.R.S.: Stochastic Processes. Courant Lecture Notes, vol. 16. AMS, Providence (2007)
67. Wheeden, R.L., Zygmund, A.: Measure and Integral. Marcel Dekker, New York (1977)
68. Zhu, H.: Dynamic programming and variational inequalities in singular stochastic control. Doctoral Dissertation, Brown University (1991)

Index

P. Azcue and N. Muler, *Stochastic Optimization in Insurance: A Dynamic Programming Approach*, SpringerBriefs in Quantitative Finance, DOI 10.1007/978-1-4939-0995-7,

S

V

MIX
Papier aus verantwortungsvollen Quellen
Paper from responsible sources
FSC® C105338

If you have any concerns about our products,
you can contact us on
ProductSafety@springernature.com

In case Publisher is established outside the EU,
the EU authorized representative is:
Springer Nature Customer Service Center GmbH
Europaplatz 3, 69115 Heidelberg, Germany

Printed by Libri Plureos GmbH
in Hamburg, Germany